Lecture Notes in Bioinformatics   13523

Subseries of Lecture Notes in Computer Science

Nicole M. Scherer ·
Raquel C. de Melo-Minardi (Eds.)

Advances in Bioinformatics and Computational Biology

15th Brazilian Symposium on Bioinformatics, BSB 2022
Buzios, Brazil, September 21–23, 2022
Proceedings

Editors
Nicole M. Scherer ⓘD
Instituto Nacional de Câncer
Rio de Janeiro, Brazil

Raquel C. de Melo-Minardi ⓘD
Universidade Federal de Minas Gerais
Belo Horizonte, Brazil

ISSN 0302-9743   ISSN 1611-3349 (electronic)
Lecture Notes in Bioinformatics
ISBN 978-3-031-21174-4   ISBN 978-3-031-21175-1 (eBook)
https://doi.org/10.1007/978-3-031-21175-1

LNCS Sublibrary: SL8 – Bioinformatics

This Springer imprint is published by the registered company Springer Nature Switzerland AG
The registered company address is: Gewerbestrasse 11, 6330 Cham, Switzerland

Preface

The Brazilian Symposium on Bioinformatics (BSB) is an international conference with a focus on bioinformatics and computational biology, organized by the special interest group in Computational Biology (CE-BioComp) of the Brazilian Computer Society (SBC). Under the coordination of the general chair, Kele Belloze from CEFET/RJ, Brazil, the 15th BSB edition was held during September 21–23, 2022, at the Atlântico Búzios Convention and Resort, in the city of Armação dos Búzios (Brazil), after two years of online-only events. BSB 2022 was co-located and jointly organized with the Brazilian Databases Symposium (SBBD 2022), which took place during September 19–13, 2022. All BSB participants could attend the SBBD activities and vice versa.

As in previous editions, BSB 2022 had an international Program Committee, which was composed of 43 members. We received a total of 50 submissions, comprising 15 full papers, eight short papers, 25 poster abstracts, and two software demonstrations. All papers were reviewed by at least three independent reviewers. After a rigorous single blind review process, a total of 17 papers (10 full papers and seven short papers) were selected to be presented in three technical sessions, and are published in this volume. The submitted abstracts and software were presented in two exciting poster sessions. In addition, BSB 2022 featured keynote talks from Yaser Hashem (Institut National de la Santé et de la Recherche Médicale, France), Peter Stadler (University of Leipzig, Germany), and Sérgio Lifschitz (PUC-Rio, Brazil).

This year we celebrated 20 years of the first Brazilian Workshop on Bioinformatics (WOB 2002), which was later renamed as the Brazilian Symposium on Bioinformatics. A round table composed of previous editions chairs, André Ponce Leon de Carvalho, Maria Emília Walter, João Setúbal, Sérgio Lifschitz, and Daniel de Oliveira, presented a retrospective of the last 20 years of BSB and discussed what we can expect to see in the next 20 years.

BSB 2022 was made possible by the dedication and work of many people and organizations. We would like to express our thanks to the general chairs of BSB and SBBD, Kele Belloze and Sérgio Lifschitz, respectively, for their tireless dedication to make this event happen, together with the members of the steering committee and the volunteers. We are also grateful to the Program Committee members, to the sponsoring institutions SBC, CEFET/RJ, FAPERJ, and CAPES, and to Springer for their continued support by agreeing to publish this proceedings volume. Last but not least, we would like to thank all authors for their time and effort in submitting their work and the invited speakers for having accepted our invitation.

September 2022

Nicole M. Scherer
Raquel C. de Melo-Minardi

Conference Topics

Big Data Analytics and High-Performance Computing in Bioinformatics and
  Computational Biology
Biochemistry
Biological Databases, Data Management, Data Integration, and Data Mining
Biological Networks
Cheminformatics and Computer-Aided Drug Design
Comparative, Structural, Functional, and Evolutionary Genomics
Computational Proteomics
Computational Systems Biology
Drug Design
Education and Training in Bioinformatics
Gene Identification, Regulation, Expression, and Post-translational Modifications
Genomics
GWAS
Machine Learning Applications in Bioinformatics
Metagenomics
Molecular Docking and Modeling
Molecular Evolution and Phylogenetics
Molecular Sequence Analysis, Motifs, and Pattern Matching
Protein Structure and Modeling
Proteomics
Single Cell Analysis
SNPs and Haplotype Analysis
Statistical Analysis of Molecular Sequences
Transcriptomics

Organization

BSB 2022 was organized by the Graduate Program in Computer Science (PPCIC) at the Federal Center for Technological Education of Rio de Janeiro (CEFET/RJ), Brazil.

General Chair

Kele Belloze   Centro Federal de Educação Tecnológica Celso
                                    Suckow da Fonseca, Brazil

Program Committee Chairs

Nicole M. Scherer   Instituto Nacional de Câncer, Brazil
Raquel C. de Melo-Minardi   Universidade Federal de Minas Gerais, Brazil

Steering Committee

Daniel de Oliveira   Universidade Federal Fluminense, Brazil
Waldeyr Mendes Silva   Instituto Federal de Goiás, Brazil
Raquel C. de Melo-Minardi   Universidade Federal de Minas Gerais, Brazil
Sérgio Lifschitz   Pontifícia Universidade Católica do Rio de
                                    Janeiro, Brazil

Program Committee

Adriano Werhli   Universidade Federal do Rio Grande, Brazil
Alejandra Medina-Rivera   Universidad Nacional Autónoma de México,
                                    Mexico
Alexandre Paschoal   Universidade Tecnológica Federal do Paraná,
                                    Brazil
Ana Carolina Guimarães   Fundação Oswaldo Cruz, Brazil
André P. L. F. de Carvalho   Universidade de São Paulo de São Carlos, Brazil
André Kashiwabara   Universidade Tecnológica Federal do Paraná,
                                    Brazil
Aristóteles Góes-Neto   Universidade Federal de Minas Gerais, Brazil
Christian Hoener zu Sieberdissen   University of Jena, Germany
Daniel de Oliveira   Universidade Federal Fluminense, Brazil
Danilo Sanches   Universidade Tecnológica Federal do Paraná,
                                    Brazil

Partner Association

Brazilian Association for Bioinformatics and Computational Biology (AB3C), Brazil

Sponsoring Institutions

Brazilian Computer Society (Sociedade Brasileira de Computação - SBC), Brazil

Foundation for Research Support in the State of Rio de Janeiro (FAPERJ), Brazil

Coordination for the Improvement of Higher Education Personnel (CAPES), Brazil

Graduate Program in Computer Science (PPCIC) at the Federal Center for
Technological Education of Rio de Janeiro (CEFET/RJ), Brazil

Contents

BDDBlast—A Memory Efficient Architecture for Pairwise Alignments

Demian Bueno de Oliveira, Alessandra Faria-Campos, and Sérgio Campos[✉]

Computer Science Department, Universidade Federal de Minas Gerais,
Belo Horizonte, MG, Brazil
scampos@dcc.ufmg.br

Abstract. BLAST, or Basic Local Alignment Search Tool, is one of
the most widely used bioinformatics tools today. However, as biological
data accumulates, its use can become a bottleneck limiting biological
analyses both in time and memory usage. In this work we propose the
use of a new data structure to re-implement BLAST. We use *Binary
Decision Diagrams* (BDDs) to store the biological sequences and optimize resources, reducing memory requirements. This new approach has
allowed us to construct the alignment of biological sequences with gains
of up to 65.7% in memory usage for the allocation of the BLAST data
structures and up to 16.3% faster search results, without altering the
BLAST algorithm or its results.

Keywords: BLAST · Basic Local Alignment Sequencing Tool · BDD ·
Binary Decision Diagrams

1 Introduction

In Bioinformatics, BLAST is one of the most well-known and widely used
tools available to search for similarities between biological sequences, such as
amino acids of proteins or nucleotides sequences of DNA. BLAST searches allow
researchers to compare an input sequence (query) against a database of sequences
(subjects) to obtain the best matches and the longest alignments for the biological sequence of study and the object(s) of comparison. The BLAST algorithm
was developed by Altschul et al. [1]. More recently Gapped BLAST [2] a more
efficient implementation has been made available. Gapped BLAST is the version
used in this work.

Currently, as more biological data are being sequenced, the demand for
more powerful platforms that are able to handle data more efficiently increases.
BLAST is an important piece of this puzzle, but it can become a very expensive
computational tool if not considered carefully.

To address this question, this work introduces **BDDBlast**, a new architecture for Gapped BLAST that uses *Binary Decision Diagrams*, or BDDs as the
major data structure. BDDs are used in different fields like formal verification,
optimization, computer-aided design (CAD), among others [3]. One of the features that makes BDDs attractive is their capacity for eliminating redundancy by
automatically eliminating paths between nodes that share common information.

© The Author(s), under exclusive license to Springer Nature Switzerland AG 2022
N. M. Scherer and R. C. de Melo-Minardi (Eds.): BSB 2022, LNBI 13523, pp. 1–13, 2022.
https://doi.org/10.1007/978-3-031-21175-1_1

BDDBlast does not change the algorithm or require special hardware. It executes the same algorithm as Gapped BLAST, and generates the same results. Preliminary experiments have shown gains up to 16.3% in search time for average sized searches and reductions up to 65.7% in memory usage for the biological data structures using real biological sequences.

While Gapped BLAST is a very efficient algorithm, several academic and commercial alternatives have been developed to increase its efficiency. They can be separated in two categories, those that use the same algorithm as Gapped BLAST, and as such are directly comparable to BDDBlast, and those that do not use the same algorithm. Some projects in the first category accelerate the original algorithm with the use of special hardware, like FPGAs (Field-Programmable Gate Array) or GPU's (Graphics Processing Units) [4,5]. PLAST [6] uses the Gapped BLAST algorithm, but it is also optimized for parallel architectures. These alternatives offer benefits with the use of parallel or cloud computing through distributed clusters of systems with hundreds of processors and large memory capacity [7]. These gains, unlike those obtained by BDDBlast, require access to restricted environments, not available to most researchers.

CS-BLAST and DELTA-BLAST [8] also use the Gapped BLAST algorithm, having as one of the main differences the use of a more sophisticated choice of score matrices. Since BDDBlast does not change the way in which these matrices are used, it is complementary to these tools, adapting the BDD data structure to be used in CS-BLAST and DELTA-BLAST could potentially multiply the savings obtained by these methods. Diamond [9] is another tool based on Gapped BLAST. It uses *double indexing* to take advantage of data locality, reducing memory usage, a similar objective as BDDBlast. It, however, does not change the data structures used, and in the same way as the previous tools, could be potentially adapted to obtain even more memory gains using BDDs.

2   System and Methods

2.1   The Gapped BLAST Algorithm

To better understand how this work explores the Gapped BLAST algorithm, two amino acids sequences will be used as an example in a walk through the main steps of a Gapped BLAST search. The first biological sequence of a search is called **Query** sequence, and represents the input data or object of study for our Gapped BLAST search. The Query is given by the user (i.e., a nucleotide or amino acid sequence). The second biological sequence is the **Subject** sequence and for this example we used a *Staphylococcus* protein sequence (GI 1004172080)[1] To make our example easier to understand, we have reduced our Query sequence to a sub-sequence of the protein sequence aforementioned:

Query: LMYKGQPMTFR
Subject: DGDTVKLMYKGQPMTFRLLLVDT

---

[1] Available at https://www.ncbi.nlm.nih.gov/protein/1004172080.

The Gapped BLAST algorithm can now be explained as follows[2]:

1 **Generate a list of words (W) from the query sequence**
Gapped BLAST first breaks the query sequence into a subset of words. Each word is then searched against the database or subject sequences. When the word is found in the subject sequence, we say there is a match. The word is then used as a seed to start a new alignment. Next, the seed is extended in both directions to expand the alignment.
2 **Compare and score the words to qualify the alignment**
The words of the query sequence are then compared to the words of the database sequences (subjects). This returns a score value, which is calculated using a scoring matrix. The scoring system evaluates the quality of an alignment. Each word scoring uses a matrix that is responsible to define which relationships are stronger and are likely to correspond to more meaningful alignments.
For proteins, Gapped BLAST uses BLOSUM62 [10] as the default substitution matrix. BLOSUM (**BLO**cks **SU**bstitution **M**atrix) is a substitution matrix (or table of values) used to score alignments between amino acids. It is one of the most widely used substitution matrices [11] and as such it provides a good testbed for BDDBlast.
In order for a word to become a seed of an initial alignment, it must also score higher than a score threshold (T). The value of T is preconfigured and can be set by the user. In our example, let's consider a threshold value of T = 15. In this case, word 1 (**LMY**), extracted from our query sequence, has aligned with the subject sequence, as seen in Fig. 1. Therefore using the scoring matrix the **LMY** sequence scores **4**, **5** and **7**, respectively. This would give us a total score of 16 at the position 7 of the subject sequence and therefore meet the requirements to become a seed of a new alignment.

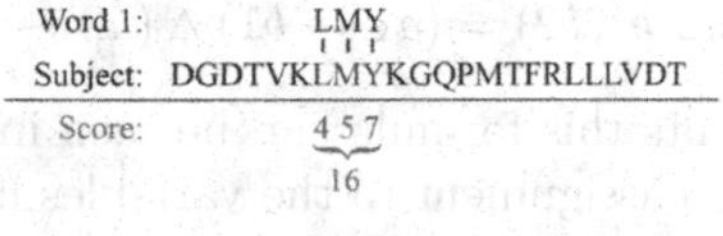

```
Word 1:          LMY
                 | | |
Subject:  DGDTVKLMYKGQPMTFRLLLVDT
Score:            4 5 7
                   16
```

Fig. 1. Words creating alignments

3 **Extend the alignment in both directions**
The seeds are then extended in both directions and an accumulated score is obtained. This expansion process stops when the total alignment score drops off by a value X when compared to the previous best score. The value of X is also defined by the user. After the expansion is complete the final alignments are called high scoring pairs (HSP).

---

[2] The value for parameters W, T and X are used to explain a simplified example, and were not used in any experiments. W and T have either default or typical values. Parameter X is mentioned for completeness, but not used in the example.

4 Preserve and Present the search results

The result of a Gapped BLAST search is a collection of HSPs that were computed during the alignment process. Figure 2 shows part of one out of the 290 HSPs that were obtained by the Gapped BLAST protein search for the example we are using. We have compared this sequence to the PDB Protein Database, using parameters word size = 3, expect value = 10, gap penalty = 11, gap extension = 1, threshold = 11, and matrix = blosum62.

```
Positives = 62/75 (83%), Gaps = 6/75

AIDGDTVKLMYKGQPMTFRLLLVDTPEFN------EK
AIDGDTVKLMYKGQ MTFRLLLVDT E        EK
AIDGDTVKLMYKGQAMTFRLLLVDTAETKHTKKGVEK
```

Fig. 2. Partial result from a Gapped BLAST search - 83% alignment

2.2   Binary Decision Diagrams (BDDs)

Binary Decision Diagrams (BDDs) were initially introduced by [12] and later explored by Bryant [13]. BDDs can be manipulated with very efficient algorithms, as demonstrated by [14].

A binary decision diagram is a rooted, directed acyclic graph with terminal and non-terminal vertices. Two terminal vertices exist, labelled 0 and 1. Each non-terminal vertex v is labeled by a variable $var(v)$ and has two successors, *zero*, followed when v is false, and *one*, followed when v is true.

To illustrate this concept and show the potential benefits of a BDD, let's consider a two bit comparator function f, given by the following formula:

$$f(a1, a2, b1, b2) = (a1 \leftrightarrow b1) \wedge (a2 \leftrightarrow b2)$$

The BDD that represents this formula can be seen in Fig. 3.

We can decide if a truth assignment to the variables in the formula satisfies f by following a path from root to leaf, following the *zero* successor if the variable is false, and the *one* successor otherwise. Applying those steps into our example, the assignment below leads to a leaf vertex labeled 0. That means the formula is false for this assignment.

$$\langle a1 \leftarrow 1, b1 \leftarrow 1, a2 \leftarrow 0, b2 \leftarrow 1 \rangle$$

BDDs explore redundancy in the representation by eliminating isomorphic and redundant subtrees, and generating extremely compact representations of boolean functions in many cases[3].

---

[3] BDDs can have a worst case exponential memory complexity in cases where the dependencies between the variables is circular or very complex. These dependencies do not typically occur in BDDBlast.

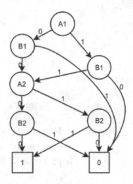

Fig. 3. Binary Decision Diagram for function f

3   Algorithm

In order to better understand BDDBlast, we will outline its main steps by exploring a hypothetical alignment result (shown below) of a protein sequence search, also extracted from the *Staphylococcus* protein example (GI 1004172080).

$$
\begin{aligned}
\text{Query} &\rightarrow \text{VDTPEFNEKYGPEASAFDKKM} \\
\text{Align} &\rightarrow \text{VDTPEFNEKYGPEASAFKKM} \\
\text{Sbjct} &\rightarrow \text{VDTPEFNEKYGPEASAFHKKM}
\end{aligned}
$$

The algorithm starts by selecting a potential seed to initiate the alignment. BDDBlast does not change how Gapped BLAST chooses the potential seed, the same algorithm is used. For this example, the word **YGP** with a score of 20 (7 + 6 + 7) was chosen to illustrate how the algorithm works. A new node is created to represent the seed word. The BDD node contains the position of the seed word in the subject sequence (value 10 in this case). This is done by associating the BDD variable in this node with the appropriate numerical value in an external data structure. For easier visualization, we show in the BDD node the word it represents, and not its numerical position.

The BDDBlast steps are described as follows:

STEP 1. The BDD construction starts with a single (root) node storing only the seed word, as illustrated in Fig. 4.

STEP 2. Once the seed word (**YGP**) has been created to represent the initial alignment between the query and subject sequences, the alignment extension process—in both directions—can start.

In this step, as the alignment is extended to the right and left sides of the root (seed word), if there is a match between the query and the subject, a new positive ($v = 1$) ramification is created. Figure 5 shows a positive path or ramification being created when the alignment is extended to the right side of the root of the query sequence given in our example.

By contrast, if there is a mismatch between the query and the subject, a new negative branch or path is created. In essence, the negative nodes store only

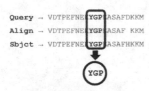

Fig. 4. Initial node representing the alignment start

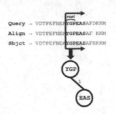

Fig. 5. Positive successor nodes, represent a match between query and subject

the subset of the subject sequence that differs from the query sequence, as all the biological sequences that match between the query and the subject (in other words, have aligned), are stored by the positive nodes. A mismatch example (the biological information that has not aligned) between the query and the subject can be seen stored in the negative node **AFH** shown in Fig. 6.

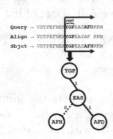

Fig. 6. Negative successor nodes represent a mismatch

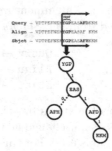

Fig. 7. Nodes generated through alignment extension

The extension process and BDD node creation will then proceed until the alignment total score drops off by a value X. For our example the complete graph generated by the extension to the right side of the root of the query sequence can be seen in Fig. 7.

STEP 3. After the graph is created, BDDBlast will eliminate the trivial paths, i.e. paths in which all nodes have only one successor, such as YGP $\rightarrow$ EAS.

A sequence of positive paths represent a subsequence that is fully matched. Therefore if there isn't any negative path between two nodes or inside a sequence of nodes, these nodes can be combined into one longer node. In our example (Fig. 5), nodes **YGP** and **EAS** are connected only by a positive path. Therefore these two nodes can be combined into a single node that represents them.

As the trivial/insignificant paths are combined into single nodes that will now be responsible to carry the information that was originally stored by multiple nodes along trivial paths, the BDDs get smaller, as can be seen in Fig. 8.

This expansion and reduction process is also applied to the negative consecutive paths sequences should they exist. In our example the left expansion

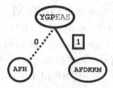

Fig. 8. BDDBlast - reduced BDD representing the right side alignment

is shown in Fig. 9, and a single node will represent this fully aligned sequence **VDTPEFNEK**, as seen in Fig. 10, once the reduction process takes place.

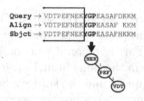

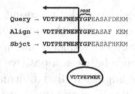

Fig. 9. Expansion to the left of the seed word

Fig. 10. Left expansion nodes combined

Reading through the Binary Decision Diagram

Once the reduced BDDs have been generated, reading its content to obtain the query, subject and therefore knowing the alignment information is straightforward. The process below shows the reading process of the BDD that represents the right side expansion of the sequences starting from the seed. The left side expansion is similar. For example, we can read the BDD in Fig. 8 as follows:

- An initial node representing the seed/root and containing the biological sequences that have aligned so far. Therefore at this point in time the query and subject information are the same as shown below:

$$Query \rightarrow YGPEAS$$
$$Align \rightarrow YGPEAS$$
$$Sbjct \rightarrow YGPEAS$$

- The negative ramification that appears after the initial node contains sequences from the subject that did not match the query. The data in this negative node goes to the subject:

$$Sbjct \rightarrow YGPEASAFH$$

in the next step when the next positive node is read, in addition to updating the query sequence, it will update the subject sequence at the offset position where it has stopped. In this case, as the negative nodes contain 3 amino acids, the offset position is marked to resume at position 4. This means that when the positive node containing the **AFDKKM** sequence is read, the subject sequence will be updated starting at the offset position 4, meaning in this case that only the **KKM** biological data is added to the subject.

– The positive node is read, updating the query sequence. It will also update the subject sequence with the additional information that has aligned after the failure point.

This takes us to the following query, subject and alignment values, under the Gapped BLAST results format:

$$\text{Query} \rightarrow \text{YGPEASAFDKKM}$$
$$\text{Align} \rightarrow \text{YGPEASAFKKM}$$
$$\text{Sbjct} \rightarrow \text{YGPEASAFHKKM}$$

The positive nodes represent the Query sequence used for the search. Overwrite the query with the negative nodes at the appropriate points and you will have the Subject sequence aligned with the query. Furthermore, if you compare the difference between a positive and negative node in the same branch, you will obtain the alignment line information between the query and the subject. Some codes are also used to represent this divergence. A gap $(-)$, a positive score $(+)$ or a non-positive score (" ") are used.

The left side expansion is performed in the same way, except that the new subsequences are appended *to the left* of the already processed subsequences. The left side must be processed and printed before the right side to generate the correct output.

4   Implementation

We have implemented a BDDBlast prototype using the NCBI Toolkit using the Gapped BLAST implementation as basis. The program was used in several tests to compare its memory usage to the original Gapped BLAST, as well as assess its performance. The tests were performed using a Lenovo laptop containing an Intel(R) Core(TM) i5-3320M CPU @2.60 GHz processor with 8 GB of RAM memory, using an IBM Edition Windows 7 64-bit v1.10.00.AG 811 operating system. Different data sets from different databases publicly available in the NCBI servers were used, including NCBI's Reference Sequence (RefSeq) database [15]. The databases used in our tests are available in the supplementary documentation.

To validate the results, 500 different tests were performed using 50 different scenarios. Each testing scenario was executed 10 times. The details of each scenario, as well as all files are available in the supplementary documentation. The tests have used Gapped BLAST's implicit default parameters (For Proteins: word size $= 3$, expect value $= 10$, gap penalty $= 11$, gap extension $= 1$, threshold $= 13$, matrix $=$ blosum62. For Nucleotides: word size $= 11$, expect value $= 10$, gap penalty $= 5$, gap extension $= 2$).

Table 1. Databases used for testing BDDBlast. Refseq is NCBI's Reference Sequence database; PDB is a protein database from PDB; NCBI is a genome chromosomes database; NT contains environmental samples; 16S is a 16S Microbial sequences database; WGS has whole-genome-shotgun sequences.

Database	Sequences	Longest seq.	Residues
Refseq Prot	2,761,826	33,467	997,236,991
Refseq Gen	141	171,031,299	3,908,663,647
PDB	82,575	5,037	20,366,134
NCBI	35	14,668,833	73,153,929
NT	3,169,069	170,452	3,103,377,759
16S	9,251	2,952	13,610,589
WGS	42,751	987,023	60,008,585

4.1 Databases and Sequences Tested

The number of sequences and sizes of the databases used are shown in Table 1. The *Staphylococcus* protein sequence, GI 1004172080, used as example in the theoretical sections of this work was also used in the majority of the tests.

To ascertain the scalability of our method we used also some large sequences to stress test the tool and verify if memory gains scale as well as the total memory usage by the tool. To this purpose we have used as query sequences Titin, or Connection, the largest human protein. We have used the DNA sequence with more than 83,000 nucleotides, as well as the protein sequence, with about 35,000 amino acids. Another set of sequences used contains 47 sequences with up to 500,000 nucleotides each, with a total of more than 6,000,000 nucleotides [16] (used with permission). The largest sequence used in this work is a fragment of the human chromosome 22 with 7,650,000 nucleotides.

Although the *Staphylococcus* protein sequence belongs to the Protein Data Bank, it was also tested against the RefSeq database so the performance of BDDBlast could be evaluated against a different data set from a genetic perspective. Tests that varied the size of the protein sequences were also conducted to verify how BDDBlast would perform from a scalability perspective.

Databases were randomly selected (othen than RefSeq) to assess if any impact in the tools performance would be noticed as database content or sizes were changed. No impact was observed.

5 Discussion

5.1 Efficient Memory Usage

A significant reduction in memory usage has been observed, as can be seen in Table 2. When the initial HSPs align perfectly, savings are expressive. As alignments start to have decreasing similarities, savings show small variations.

Table 2. Chart showing gains as searches are scaled up - Memory usage - *Staphylococcus* and PDB

Query size (residues)	Memory used (bytes)		
	BLAST	BDDBlast	Mem. gain (%)
100	2,942,400	1,401,408	52.37%
149	6,002,688	2,072,704	65.47%
304	14,543,616	4,989,760	65.69%
507	1,686,144	810,048	51.96%
1000	179,427,840	90,932,480	49.32%
2001	82,115,712	41,035,840	50.03%
4011	70,974,528	37,405,248	47.30%
5005	7,734,912	3,522,112	54.46%

Tests involving a biological sequence of the *Staphylococcus* protein against the PDB (Protein Data Bank) have provided gains of up to 65.7% in memory usage to represent the biological data structures.

Finally, the *Staphylococcus* sequences were tested against the RefSeq database. This test evaluates how the BDDBlast algorithm would perform when tested against a database source different from the protein source. It is expected that the similarities will be smaller in this case. The memory gains have dropped in this scenario but are still very positive, showing an average gain of more than 58% when compared to Gapped BLAST.

Table 3. Memory gains for very large sequences.

Sequence	DB	Seq. size	BLAST mem (Mbytes)	BDDBlast mem (Mbytes)	Gains	BDDBlast memory allocated
Titin gen.	RF Gen	81,940	11.8	4.4	62.91%	674 MB
Titin prot.	RF Prot	34,993	3,069	1,635	46.73%	703 MB
Corumba90k	NT	6,276,248	187	76	59.41%	716 MB
CHR22-150k	RF Gen	7,650,064	1,863.6	727.5	60.96%	1,071 MB
CHR22-150k	NCBI	7,650,064	52,212	20,314.5	61.09%	1,982 MB

We have also tested the proposed methodology in very large sequences to determine if savings scale up. We used the same databases but used three different sequence sets of increasing size. The results of these comparisons are shown in Table 3, showing consistent memory gains in the order of 60% (with protein comparisons showing slightly smaller gains).

5.2   Faster Execution Time

In addition to memory gains, CPU execution time speedups have also been observed with gains of up to 16.3% in faster search results. The execution

Table 4. Table showing gain levels as searches are scaled up - CPU time - *Staphylococcus* and PDB

Query size (residues)	CPU (time in seconds)		
	BLAST	BDDBlast	Time gain (%)
100	10,185	9,244	9.24%
149	19,870	16,987	14.51%
304	22,791	19,896	12.70%
507	46,991	43,071	8.19%
1000	69,980	62,830	10.22%
2001	123,766	113,665	8.16%
4011	165,780	148,539	10.40%
5005	192,462	179,264	6.86%

times reported correspond to the complete execution of the Gapped BLAST or BDDBlast programs. As previously noticed in the memory gains, the CPU savings are also consistent for sequences of different sizes, but not for very large sequences, as will be seen. Table 4 shows a comparison table between Gapped BLAST and BDDBlast for CPU execution time with different search sizes.

An additional test involved the RefSeq database, also used for the Memory testing. As previously mentioned, this test aimed to check a different database from the protein's origin database. As the alignment levels were reduced, the CPU execution time showed a slight drop. However even in this scenario BDDBlast was able to outperform the original Gapped BLAST architecture, with gains averaging about 10%.

For the very large sequences tests time results show a slight *slowdown* when using BDDBlast, averaging 8.84%. It is, however, much less significant than the memory usage gains. Further research is needed to optimize running times in these cases.

6   Conclusion

This work proposes a version of Gapped BLAST that incorporates the benefits of Binary Decision Diagrams. The new BDDBlast architecture allows the construction of more efficient pairwise alignments for biological sequences, resulting in memory and time savings. The performance improvements have shown to be consistent and therefore potentially scalable.

The relevance of our result is that we have shown a novel approach to manage and store the Gapped BLAST alignments, under a known data structure but that had never been tried before. This work opens the path towards a new optimization discussion for the Basic Local Alignment Tool through the use of the BDDs.

Future work includes applying BDDs to Gapped BLAST implementations that optimizations that modify substitution matrices or the order in which operations are executed. It is expected that gains can be obtained in those cases as well, *in addition* to the gains of these tools.

Another area of future exploration is the use of BDDBlast in comparisons between full databases, i.e., comparing multiple input sequences to a database. While intuitively the behavior can be seen as the addition of the time taken by each comparison, it is possible to share BDD nodes between different comparisons, obtaining even larger gains. Implementing this idea is left for future work.

Acknowledgments. The authors would like to thanks Ronnie Alves and Kleber de Souza with assistance with the Corumba sequences. This research was supported by in part by CNPq, CAPES and FAPEMIG, Brazilian Funding Agencies.

Availability. BDDBlast and the sequences used in this paper can be found at: http://www.luar.dcc.ufmg.br/bddblast/.

References

1. Altschul, S.F., et al.: Basic local alignment search tool. J. Mol. Biol. **215**(3), 403–410 (1990)
2. Altschul, S.F., et al.: Gapped blast and psi-blast: a new generation of protein database search programs. Nucleic Acids Res. **25**(17), 3389 (1997)
3. Clarke, E.M., Grumberg, O., Peled, D.: Model Checking. MIT Press, Cambridge (1999)
4. Kasap, S., Benkrid, K., Liu, Y.: Design and implementation of an FPGA-based core for gapped blast sequence alignment. Eng. Lett. **16** (2008)
5. Ye, W., et al.: H-blast: a fast protein sequence alignment toolkit on heterogeneous computers with GPUs. Bioinformatics **33**(8), 1130–1138 (2017)
6. Nguyen, V., Lavenier, D.: PLAST: parallel local alignment search tool for database comparison. BMC Bioinform. **10**(329), 1–13 (2009)
7. Fahim, S.: Comparative analysis of protein alignment algorithms in parallel environment using CUDA. Ph.D. dissertation, BRAC University (2016)
8. Boratyn, G.M., Schäffer, A.A., Agarwala, R., et al.: Domain enhanced lookup time accelerated blast. Biol. Direct **7**(12), 1–14 (2012). https://doi.org/10.1186/1745-6150-7-12
9. Buchfink, B., Xie, C., Huson, D.: Fast and sensitive protein alignment using diamond. Nat. Methods **12**, 59–60 (2015)
10. Eddy, S.R.: Where did the blosum62 alignment score matrix come from? Nat. Biotechnol. **22**(8), 1035–1036 (2004)
11. Hess, M., et al.: Addressing inaccuracies in BLOSUM computation improves homology search performance. BMC Bioinform. **17**(1), 1–10 (2016). https://doi.org/10.1186/s12859-016-1060-3
12. Lee, C.-Y.: Representation of switching circuits by binary-decision programs. Bell Labs Tech. J. **38**(4), 985–999 (1959)

13. Bryant, R.E.: Graph-based algorithms for Boolean function manipulation. Comput. IEEE Trans. **100**(8), 677–691 (1986)
14. Brace, K.S., Rudell, R.L., Bryant, R.E.: Efficient implementation of a BDD package. In: 27th ACM/IEEE Design Automation conference, p. 40. ACM (1991)
15. Pruitt, K.D., Tatusova, T., Maglott, D.R.: NCBI reference sequences (RefSeq): a curated non-redundant sequence database of genomes, transcripts and proteins. Nucleic Acids Res. **35**(suppl 1), D61–D65 (2006)
16. Gastauer, M., et al.: Ti - a metagenomic survey of soil microbial communities along a rehabilitation chronosequence after iron ore mining. Sci. Data **6**, 1–10 (2019)

Scientific Workflow Interactions: An Application to Cancer Gene Identification

Diogo Munaro Vieira[1]([⊠])[iD], Alexandre Heine[1][iD],
Elvismary Molina de Armas[1][iD], Cristóvão Antunes de Lanna[2][iD],
Mariana Boroni[2][iD], and Sérgio Lifschitz[1][iD]

[1] Laboratório BioBD - Departamento de Informática, PUC-Rio,
Rio de Janeiro , Brazil
{dvieira,aheine,earmas,sergio}@inf.puc-rio.br
[2] Laboratório de Bioinformática e Biologia Computacional, INCA,
Rio de Janeiro, Brazil
{cristovao.lanna,mariana.boroni}@inca.gov.br

Abstract. Reproducibility, resilience, and large-scale data processing have become fundamental for developing scientific research, particularly in bioinformatics. One may consider the use of Scientific Workflow Management Systems (SWfMS) to address these topics. However, user interactivity during the execution of workflows, especially with a preliminary result generated by an inner workflow task, is still an issue. We present in this paper an architecture that meets the interactive requirements of these systems, allowing the development of a flexible layer for end users to interact directly with SWfMS. Besides presenting our software solution, we show an application in the context of cancer gene identification for drug design.

Keywords: Bioinformatics · Workflows · Human-in-the-loop · Interactions · Cancer

1 Introduction

Reproducibility is an essential aspect of science and has been increasingly demanded to publish scientific works. The ease of reproducibility increases confidence in the scientific environment and allows the development of new research by additional collaborators [13].

In addition to reproducibility, by having processes and data flow standardization, data provenance can also be managed [5]. This is a very relevant point, particularly with the beginning of the General Data Protection Regulation (GDPR) and other similar initiatives around the world focused on governance and data security [1].

A scientific workflow is a chain of processes that allows the automation of tasks more straightforwardly using the necessary computational resources

Supported by the Brazilian Science and Technology Ministery and by CAPES Funding Agency.

[3,15]. Scientific Workflows Management Systems (SWfMS) improve experiment reproducibility, data provenance, and governance of scientific processes.

Interactive workflows allows people with no computer ability to execute scientific tasks [15] and improve workflow execution with additional human knowledge. This makes scientific development more accessible and will enable researchers from various branches of science to reproduce published studies or even carry out new studies by reusing other previously established workflows [2,11].

The SWfMS needs a more intuitive interface for the use case in question to achieve greater interactivity. Not all problems will be better illustrated as generic workflows as mentioned in [2] as we must interact between one workflow stage and another. The concept of Human-in-the-loop (HITL) may be applied to workflows [14,20]. The most common interaction in a workflow is between the output of one step and the start of the other. Some examples of workflow systems implemented for specific purposes allow interactivity [10,16], but not applied to cancer genes identification.

This paper proposes an architecture that implements this interaction concept only to detect cancer gene hubs as a study case. Cancer is a multifactorial and highly heterogeneous disease affecting the response to therapy and patient prognosis [4]. Efforts in improving treatment in oncology have centered around grouping tumors into molecular subtypes based on their characteristics and developing specific therapies targeting each subtype's main players [23]. An interactive workflow allows the user to choose whether to stratify analyzed samples into groups representing molecular subtypes or make other choices based on their expert knowledge. New applications will be contemplated in the future.

2 Methods

During workflow development, it is first necessary to strategically organize its steps, as well as the interaction between each of them. After that, we need an analysis of each SWfMS to choose the most suitable for your particular workflow.

2.1 Workflow Abstraction

The importance of an abstract workflow goes beyond the organization of its stages. The standardization also favors its use to scale each process in a simpler way [6].

In this work, we develop the abstract workflow. It comprises three main steps: Data preparation, Network Analysis, and Validation. To analyze the interaction points, we also expanded the *Network Analysis* showing its several steps that can be precisely one or several processes as illustrated in Fig. 1. After that, it is possible to explore some interaction points in each stage and think about how the user will do these interactions.

In the development of this work, we explored two main points of interaction. First, when the gene expression table emerges from *Data Preparation*, the

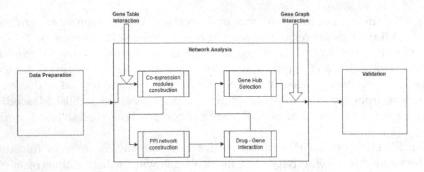

Fig. 1. Abstract workflow with *Network Analysis* component expanded. A gene expression table is provided by the user and undergoes grouping based on expression values, leading to the construction of *Co-expression modules*. Then, each module is provided for protein-protein interactions (PPI) search and modules are converted into *PPI networks*. *Drug-gene interactions* are also evaluated for each component of the network. Finally, *gene hubs* are selected for each PPI network based on the network's topology. User interaction for gene table and graphs are marked.

user should be able to remove genes according to some stipulated business rule, such as genes with low expression across all samples or with low variability. The second point is at the end of the processing of the *Network Analysis* component at the end of the processing, over the output of the component when the genes have undergone grouping by co-expression analysis, network building using protein-protein interactions based on each co-expression module, identification of network hubs, and drugs associated with network components. At this point, interaction graphs allow the user to choose the best ones to continue the workflow by selecting potential cancer drivers from the hubs list or good candidates for drug repositioning, for example. These points are illustrated in Fig. 1.

The interaction ends up not being a functionality nor a responsibility of the SWfMS, so we chose an interactive web interface made with Django [7] and React JS [8].

2.2   Choosing SWfMS

We need to choose an SWfMS that would meet the demands for user interactions, as well as other project requirements.

In our case, the workflow needs (i) **external integration** (API or CLI) for the interactions, (ii) **task cache** because tasks have a high computational cost; (iii) **parallelism**, as the same operation will be processed in multiple genes; and (iv) **version control**, because if the workflow is changed, users need to know that particular version that was executed for future executions.

2.3   Interactions

The *Prefect* framework [19] met all the requirements and was chosen to develop this research work. It is written in Python [24] and contains many features to

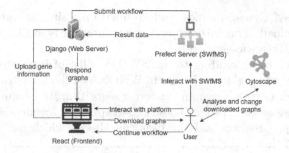

Fig. 2. HITL with workflow using Cytoscape for human analysis.

avoid *Negative Engineering* [17], which is the development of code to work around problems that a workflow manager should solve for their code to run without needing too many adaptations.

3   Results

The architecture, user interactions integration and results of this interactions with the workflow build on top of Prefect are shown in the current section.

3.1   Architecture

The architecture design is composed by: a Frontend for user interaction; a Web Server that controls access and make Frontend interaction with SWfMS; and the SWfMS that process tasks and is available for user debugging. The interaction between the user and all components of this architecture is illustrated in Fig. 2.

Prefect's architecture is composed of the interaction between several services[1] In this architecture, the most important part is the API that we can interact with the workflow and build an interactive experience.

The Django framework was chosen as the Server-Side development platform for interactions with the workflow. As an interactivity technology, React was chosen because it is simpler to port to applications and because it already has some visual components ready on the internet. Along with React, the Cytoscape [21, 22] pattern was used for graphs interchange during HITL interactions.

The user can interact through the SWfMS UI just by running the workflow or using the interface that allows interactivity between the workflow steps. If the option is to use the interactive interface, this interface communicates directly with the server in Django (Web Server) that asks the Prefect (SWfMS) to execute that step of the workflow. After execution, the SWfMS returns the response with the processing data and the resulting graph clusters to the Web Server. Then, these graph clusters are saved on a database and it is displayed a download option to the user. With this option, the clusters are retrieved from the database,

---

[1] Prefect architecture: https://docs.prefect.io/orchestration/server/architecture.html.

saved on separated Cytoscape files, and grouped on a single compacted file for the user to download. This interactive implementation is available at https://github.com/BioBD/sgwfc-gene-web.

The researcher starts sending to the SWfMS a list of interested genes. The SWfMS starts clustering modules with WGCNA [12] and enrich data with StringDB [9], later Girvan-Newman clustering algorithm [18] is applied to obtain genes communities as sub-graphs. These sub-graphs are downloaded to be analysed in Cytoscape program and the user can choose which one will continue the workflow. All theses workflow steps were implemented as tasks and now are available at https://github.com/BioBD/sgwfc-gene-python.

4 Conclusions

In this work, we show the feasibility of developing a workflow with interactive inputs and outputs in each task that works with gene communities identification with the SWfMS Prefect. This architecture enables HITL that empower users to contribute with their knowledge.

In future work, it is necessary to thoroughly implement the workflow, developing the interactive steps according to the project's needs and designing a workflow that is fully interactive and adaptable to the user's needs. Another future work is that workflow can learn from previous submissions and suggest better sub-graphs to new users.

References

1. Albrecht, J.: How the GDPR will change the world. Eur. Data Prot. Law Rev. **2**(3), 287–289 (2017). https://doi.org/10.21552/edpl/2016/3/4
2. Altintas, I., Berkley, C., Jaeger, E., Jones, M., Ludäscher, B., Mock, S.: Kepler: an extensible system for design and execution of scientific workflows. In: Proceedings of the International Conference on Scientific and Statistical Database Management, SSDBM, vol. 16, pp. 423–424 (2004). https://doi.org/10.1109/ssdm.2004.1311241
3. Barker, A., van Hemert, J.: Scientific workflow: a survey and research directions. In: Wyrzykowski, R., Dongarra, J., Karczewski, K., Wasniewski, J. (eds.) PPAM 2007. LNCS, vol. 4967, pp. 746–753. Springer, Heidelberg (2008). https://doi.org/10.1007/978-3-540-68111-3_78
4. Dagogo-Jack, I., Shaw, A.T.: Tumour heterogeneity and resistance to cancer therapies. Nat. Rev. Clin. Oncol. **15**, 81–94 (2018). https://doi.org/10.1038/nrclinonc.2017.166
5. Davidson, S.B., et al.: Provenance in scientific workflow systems. IEEE Data Eng. Bull. **30**(4), 44–50 (2007)
6. DeelmanDeelman, E., et al.: Mapping abstract complex workflows onto grid environments. J. Grid Comput. **1**(1), 25–39 (2003). https://doi.org/10.1023/A:1024000426962
7. Django Software Foundation: Django. https://djangoproject.com
8. Facebook: React js. https://reactjs.org
9. Franceschini, A., Franceschini, M.A., RUnit, S., biocViews Network, B.: Package 'STRINGdb' (2015)

10. Guan, Z., et al.: Grid-flow: a grid-enabled scientific workflow system with a petri-net-based interface. Concurrency Comput. Pract. Experience **18**(10), 1115–1140 (2006). https://doi.org/10.1002/cpe.988
11. Keefe, D.F.: Integrating visualization and interaction research to improve scientific workflows. IEEE Comput. Graph. Appl. **30**(2), 8–13 (2010). https://doi.org/10.1109/MCG.2010.30
12. Langfelder, P., Horvath, S.: Wgcna: an r package for weighted correlation network analysis. BMC Bioinf. **9**(1), 1–13 (2008)
13. Laraway, S., Snycerski, S., Pradhan, S., Huitema, B.E.: An overview of scientific repro-ducibility: consideration of relevant issues for behavior science/analysis. Perspect. Behav. Sci. **42**(1), 33–57 (2019). https://doi.org/10.1007/s40614-019-00193-3
14. Liu, J., Wilson, A., Gunning, D.: Workflow-based human-in-the-loop data ana-lytics. In: ACM International Conference Proceeding Series, pp. 49–52 (2014). https://doi.org/10.1145/2609876.2609888
15. McPhillips, T., Bowers, S., Zinn, D., Ludäscher, B.: Scientific workflow design for mere mortals. Futur. Gener. Comput. Syst. **25**(5), 541–551 (2009). https://doi.org/10.1016/j.future.2008.06.013
16. Mohammed, Y., et al.: PeptidePicker: a scientific workflow with web interface for selecting appropriate peptides for targeted proteomics experiments. J. Proteomics **106**, 151–161 (2014). https://doi.org/10.1016/j.jprot.2014.04.018
17. Newsletter, F.: What Is Negative Engineering? (2022). https://future.com/negative-engineering-and-the-art-of-failing-successfully
18. Pinney, J.W., Westhead, D.R.: Betweenness-based decomposition methods for social and biological networks. Interdis. Stat. Bioinf. **25**, 87–90 (2006)
19. Prefect Technologies Inc: Prefect. https://www.prefect.io
20. Rahman, S., Kandogan, E.: Characterizing practices, limitations, and opportunities related to text information extraction workflows: a human-in-the-loop perspective. In: Association for Computing Machinery (ACM), pp. 1–15 (2022). https://doi.org/10.1145/3491102.3502068
21. Shannon, P., et al.: Cytoscape: a software environment for integrated models of biomolecular interaction networks. Genome Res. **13**(11), 2498–2504 (2003)
22. Smoot, M.E., Ono, K., Ruscheinski, J., Wang, P.L., Ideker, T.: Cytoscape 2.8: new features for data integration and network visualization. Bioinformatics 27(3), 431–432 (2010). https://doi.org/10.1093/bioinformatics/btq675
23. Tsimberidou, A.M., Fountzilas, E., Nikanjam, M., Kurzrock, R.: Review of preci-sion cancer medicine: evolution of the treatment paradigm. Cancer Treat. Rev. **86**, 102019 (2020). https://doi.org/10.1016/j.ctrv.2020.102019
24. Van Rossum, G., Drake, F.L.: Python 3 Reference Manual. CreateSpace, Scotts Valley (2009)

Accuracy of RNA Structure Prediction Depends on the Pseudoknot Grammar

Dustyn Eggers[1], Christian Höner zu Siederdissen[1,2]🆔,
and Peter F. Stadler[1,3,4,5,6(✉)]🆔

[1] Bioinformatics Group, Department of Computer Science,
and Interdisciplinary Center for Bioinformatics, Universität Leipzig,
Härtelstrasse 16-18, 04107 Leipzig, Germany
studla@bioinf.uni-leipzig.de
[2] Theoretical Bioinformatics Lab, Bioinformatics/High-Throughput Analysis,
Faculty of Mathematics and Computer Science,
Friedrich Schiller University Jena, Leutragraben 1, 07743 Jena, Germany
christian.hoener.zu.siederdissen@uni-jena.de
[3] Max Planck Institute for Mathematics in the Sciences, Leipzig, Germany
[4] Institute for Theoretical Chemistry, University of Vienna, Vienna, Austria
[5] Facultad de Ciencias, Universidad Nacional de Colombia, Bogotá, Colombia
[6] Santa Fe Institute, Santa Fe, NM, USA

Abstract. Most models for pseudoknotted RNA structures can be described by multi-context free grammars (MCFGs) and thus are amenable to dynamic programming algorithms. They differ strongly in their definition of admissible structures and thus the search space over which structures are optimized. The accuracy of structure prediction can be expected to depend on choice of the MCFG: models that are too inclusive likely over-predict pseudoknots, while restrictive models by their definition already exclude more complex pseudoknotted structures. A systematic analysis of the impact of the grammar, however, is difficult since available implementations use incomparable energy parameters. We show here that Algebraic Dynamic Programming over MCFGs naturally disentangles energy models (as specified by the evaluation algebra) and the definition of search space defined by a MCFG. Preliminary computational experiments indicate that the choice of the grammar has an important impact already for short RNA sequences.

Keywords: RNA Pseudoknots · Multi-context free grammar · Algebraic dynamic programming · ADPfusion

This work was funded by the German DFG Collaborative Research Centre AquaDiva (CRC 1076 AquaDiva), the German state of Thuringia via the Thüringer Aufbaubank (2021 FGI 0009), the Carl-Zeiss-Stiftung within the program Scientific Breakthroughs in Artificial Intelligence (project "Interactive Inference"), and the German Federal Ministry of Education and Research (BMBF 031L0164C "RNAProNet").

N. M. Scherer and R. C. de Melo-Minardi (Eds.): BSB 2022, LNBI 13523, pp. 20–31, 2022.
https://doi.org/10.1007/978-3-031-21175-1_3

1 Introduction

RNA secondary structure without pseudoknots can be predicted efficiently by means of dynamic programming using a well-established standard energy model. Pseudoknots, however, play an important role in RNA function, contributing in particular to the regulation of translation and splicing, and ribosomal frameshifts [3,24]. With `pseudobase` there is a dedicated repository of biologically relevant RNA pseudoknots [26]. The RNA folding problem for general pseudoknotted structures and energy models that depend on stacked base pairs can be formally stated as follows:

General RNA Folding
Input: An ordered sequence of vertices $(x_1, \ldots, x_k)$, weights $\omega(i, j; k, l)$ that are non-positive only for "stacked edges", i.e., if $k = i+1$ and $l = j-1$, and a bound E.
Question: Is there a matching M, i.e., a set of edges such that each vertex is incident to at most one edge, with $f(M) := \sum_{\{i,j\},\{k,l\}\in M} \omega(i, j; k, l) \le E$?

The GENERAL RNA FOLDING problem is known to be NP-complete if arbitrary stacking energies $\omega(.)$ can be used [1]. It remains NP complete in the unweighted case, i.e., for $\omega(i, j; i+1, j-1) = 1$ if and only if $x_i x_j$ and $x_{i+1} x_{j-1}$ are Watson-Crick base pairs [10]. Additional hardness results can be found in [20].

Several research groups proposed dynamic programming algorithms that solve the corresponding folding problem for certain restricted classes of matchings M with restrictions on the patterns of crossing edges $\{i, j\}$ and $\{k, l\}$ with $i < j < j < l$) forming the pseudoknots. These algorithms differ drastically in their definition of admissible pseudoknot types and thus in the extent of the search space, see [4,12] for an overview. The performance of the different algorithms is difficult to compare because they typically employ different parametrizations of the energy model and thus already differ in their prediction of structure without pseudoknots. It is hard to decide, therefore, whether differences in the prediction accuracy are the consequence of better energy parameters for the stems and loops of pseudoknot-free parts of the structure, or whether they have to be attributed to the pseudoknots. It has remained an open question, therefore, whether the choice of the search space has an important impact, and whether there is an optimal pseudoknot model that is sufficiently inclusive to cover the known structures but rules out structures that are impossible or unlikely to be realized at all.

In this contribution we consider re-implementations of different pseudoknot models in a common framework. This allows us, in particular, to ensure that all knot-free structures and substructures are handled identically. Furthermore, it makes it possible to assign the same energy contributions to matching types of pseudoknots. Dynamic programming (DP) algorithms are commonly defined as recursion relations that iteratively fill memo-tables. These tables are often indexed by complex structures that make the implementation of DP recursions a tedious and error prone task [6]. The theory of Algebraic Dynamic Programming

(ADP) [7] addresses this issue for a restricted class of DP algorithms for which (i) generation of the state space, (ii) scoring of states, and (iii) selection of desired solution can be separated completely. ADP is therefore the ideal framework for our endeavor, although there are attractive alternative abstract formalisms, such as "super-grammars" [18], forward-hypergraphs as an alternative description of dependencies [13] and inverse coupled rewrite systems (ICORES) [8].

2   Algebraic Dynamic Programming and ADPfusion

ADP utilizes a *grammar* to specify the state space and thus the structure of the recursion without any explicit reference to indices. The original setting of ADP are context-free languages, and thus productions of the form $A \rightarrow \alpha$, where A is a non-terminal and α is an arbitrary expression formed from terminals and non-terminals [7]. More recently, the formalism was extended to so-called multi-context-free grammars [16]. The main difference is that non-terminals may now be multi-dimensional, corresponding to non-overlapping sub-objects that are parsed independently. For both CFGs and MCFGs, each production determines a partition of the non-terminal on the l.h.s. Interpreting each non-terminal on the r.h.s. as a parser alleviates the need to specify indices explicitly. For instance, the simple production $S \rightarrow [S]S$ corresponds to the recursion relation $S_{ij} \mathrel{+}= \sum_{k=i+1}^{j} c_i S_{i+1,k-1} \bar{c}_k S_{k+1,j}$, (with the convention that the empty parse $S_{k,k-1} = 1$ serves as neutral element), where the sequence interval $[i,j]$ on which a structure "lives" is indicated by the index pair S_{ij}. The terminals c and $\bar{c}$ together signify a base pair.

In ADP, each production is interpreted by an *evaluation algebra*. Productions as grammatical objects are linked to the evaluation algebra via a common (type) signature. To understand this connection, consider the grammar $\{S \rightarrow cScS\}$, where the brackets '[' and ']' have been generalized to accommodate any particular character. The r.h.s. of the single rule of this grammar has the following "type": $c \times x \times c \times x$, while the l.h.s. holds objects that evaluate to x. The full type of the rule then is $c \times x \times c \times x \rightarrow x$. This type signature provides a constraint for both the grammar and the evaluation of parses of inputs. The terminal types c indicate that single characters are to be matched upon, while x indicates *not only* that the parse has to continue recursively *but also* that each recursive parse can immediately be replaced by a value of type x *(by means of memoization)*, for example a locally optimal score. This finally points to the structure of the evaluation algebra. An evaluation algebra is devoid of any structural notion. Instead it only contains functions that *interpret* each parse and immediately replace it by a value. In the example above, for each i, k, j, the parse $c_i S_{i+1,k-1} \bar{c}_k S_{k+1,j}$ is evaluated by a function of type, say, $\texttt{Char} \times \texttt{Int} \times \texttt{Char} \times \texttt{Int} \rightarrow \texttt{Int}$, which is either $S_{i+1,k-1} + S_{k+1,j} + 1$, if c_i and $\bar{c}_k$ are pairing, or $-\infty$.

The same grammar thus can be used to minimize scores, compute partition functions, density of states, or enumerate a fixed number of sub-optimal solutions by simply employing a different evaluation algebra. In addition, not only is each algebra comparátively simple, the notion of product operations on algebras allows for easy combination of different algebras to calculate diverse and complex questions over grammars [25].

We employ a particular variant of the idea of ADP, namely ADPfusion [21]. This framework performs in-depth program fusion during compilation, which effectively turns a very high-level *declarative* description of a dynamic programming into tight loops that operate directly on flat memory. Authors of dynamic programs may freely mix different types of grammars, which can operate on diverse and heterogeneous index spaces [23] while still producing the desired, efficient loops that are required for dynamic programs that are asymptotically costly.

The latter property is very useful for the type of grammars we are interested in here. Multiple Context-Free Grammars (MCFGs) [19] are a particular type of weakly context-sensitive grammar that, in contrast to the general case, employ in their rewrite rules only total functions that concatenate constant strings and components of their arguments. As a consequence MCFGs admit polynomial-time parsing, i.e., the membership of word w of length n in a language generated by an MCFG G can be determined in $O(n^{c(G)})$, with a constant $c(G)$ depending only on the grammar.

Each MCFG còntains rules that conform to the canonical pseudoknot-free structures – and thus substrings that are juxtaposed – and rules over substrings that contain "holes" and are interleaved with each other. The latter are represented by higher-dimensional index objects. MCFGs therefore operate on non-terminals that have an interpretation as tuples of strings over an alphabet A – rather than strings as in the case of CFGs. Due to space constraints, we cannot give a formal presentation of MCFGs here and instead refer to [16]. Instead, we use the minimal pseudoknot model of GenussFold [16] as a means of explaining the notations at an operational level. Consider the following productions:

$$S \to \epsilon \mid \bullet S \mid [S] S \mid A_1 B_1 A_2 B_2$$

$$\begin{pmatrix} A_1 \\ A_2 \end{pmatrix} \to \begin{pmatrix} S[A_1] \\ A_2 S] \end{pmatrix} \mid \begin{pmatrix} \epsilon \\ \epsilon \end{pmatrix} \qquad \begin{pmatrix} B_1 \\ B_2 \end{pmatrix} \to \begin{pmatrix} S[B_1] \\ B_2 S] \end{pmatrix} \mid \begin{pmatrix} \epsilon \\ \epsilon \end{pmatrix} \qquad (1)$$

In addition to the terminals ϵ, $\bullet$, [,] , which refer to the empty string, a single unpaired nucleotide and base pair, this MCFG uses three non-terminals: the one-dimensional nonterminal S represents arbitrary structures. The two-dimensional terminals $\begin{pmatrix} A_1 \\ A_2 \end{pmatrix}$ and $\begin{pmatrix} B_1 \\ B_2 \end{pmatrix}$ describe the two interleaved interacting parts of an H-type pseudoknot. Note that any one-dimensional index is represented by the tuple (i,j) with $i \le j$ to fully identify a substring. A simple example of a successful parse of the string [{]} is given in Fig. 1.

In [16,22] we introduced a domain specific langauge (DSL) that makes it fairly convenient to write productions with 2-dimensional non-terminals. Here, we employ the same idea. First, the l.h.s. is "reformatted" such that the com-

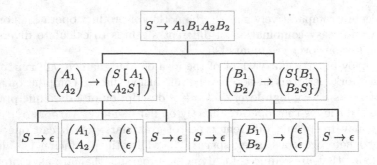

Fig. 1. Parse tree of the string [{]}. Compared to Eq. 1, the terminal symbols for the case $\begin{pmatrix}B_1\\B_2\end{pmatrix}$ have been replaced with the symbols {,} to emphasize the two terminals forming base pairs: [pairs with], while { pairs with }. The rule $S \rightarrow A_1 B_1 A_2 B_2$ splits the string into four (possibly empty) substrings, say, $[_{1,1}\{_{2,2},]_{3,3},\}_{4,4}$. The two-dimensional rule $\begin{pmatrix}A_1\\A_2\end{pmatrix}_{3,3}^{1,1}$ then operates on *pairs* of indices simultaneously, while the construction guarantees that only *legal* parses are derivable, i.e., the parse over $(1,1),(3,3)$ for A_1 and A_2.

ponents of the 2-dimensional non-terminal are aligned:

$$\begin{pmatrix}S\,[\,A_1\\A_2S\,]\end{pmatrix} \rightsquigarrow \begin{pmatrix}S\;[\;A_1\;-\;-\\-\;-\;A_2\;S\;]\end{pmatrix}$$

and then each column is transposed into a tuple to obtain a linear text

`[S,-] [nt,-] <A,A> [-,S] [-,nt]`

The "gap symbols" – are used to specify whether one-dimensional terminals and non-terminals `nt` and `S` refer to the first or second dimension. The DSL also suppresses the indices of the components of two dimensional non-terminals. One thus simply writes

```
S        -> hpk <<< A B A B
<A,A> -> pka <<< [S,-] [nt,-] <A,A> [-,S] [-,nt]
<B,B> -> pkb <<< [S,-] [nt,-] <B,B> [-,S] [-,nt]
```

following as far as possible the notational convention of other ADP implementations [7].

Dynamic programming can be used to answer more complicated questions than the computation of maximum likelihood (or more generally score-optimal) solutions. One important class of problems concerns the relative likelihood with which a substructure occurs, weighted by its likelihood. This question, which also appears e.g. in certain algorithms for parameter fitting, requires a combination of *inside and outside* algorithms. These two algorithms describe the same search space. While the inside algorithm operates bottom-up, the corresponding outside algorithm traverses the search space in top-down order. Traditionally, the outside algorithm is carefully constructed by hand to correctly match in all cases and generate exactly the same probabilities (or scores). It is possible to fully automate this construction [23] along with the required conversions of the each evaluation algebra. While not shown here, this automated construction is available in ADPfusion and thus for all grammars we consider here. This yields, for instance, algorithms to compute Boltzmann-weighted base pairing probabilities for the different classes of pseudoknotted structures.

3 Pseudoknot Grammars

The context-free grammar describing the folding algorithms for pseudoknot-free structures as implemented e.g. in the ViennaRNA package [9] can be written in the following form

$$
S \to \epsilon \mid \bullet S \mid BS
$$
$$
B \to cr\bar{c} \mid crBr'\bar{c} \mid cMM'\bar{c} \tag{2}
$$
$$
M \to rB \mid MB \mid M \bullet \qquad M' \to B \mid M'\bullet
$$

The non-terminals denote an arbitrary structure (S), a structure enclosed by a base pair (B), a component of a multiloop with at least one base pair inside (M), and a multiloop component whose initial base is paired (M'). The grammar conforms to the standard energy model for RNA secondary structures [27], which distinguishes hairpin-loops, interior loops (including base pairs) with a single enclosed base pair, and multiloops with two or more enclosed pairs. The terminals $\bullet$, and $c, \bar{c}$ denote an unpaired base and base pair, respectively. In addition, we write r for a region without base pairs of length at least 1 and r, r' for a pair of regions of total length at least 1. The last two lines implement the *loop decomposition*, i.e., distinguishes hairpin, interior, and multibranch loops and decomposes multiloops to support and energies that are linear in the number of unpaired bases and the number emanating stems.

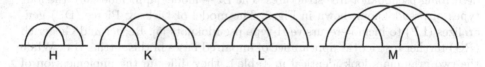

Fig. 2. The four types of pseudoknots with topological genus 1 [15] correspond to H-type pseudoknots (H), kissing hairpins (K) and two types of rare, more complex pseudoknots. The four types of pseudoknots correspond to the four alternatives in the LP^+ and RN grammars.

Many of the competing models of pseudoknots are compared in terms of their MCFG grammars and languages in [12]. Table 1 summarizes the subset considered in the contributions: The RE (Rivas & Eddy) model specifies the most

Table 1. Overview of the Pseudoknot Grammars, adapted from [4] and [12].

LP	$S \to \epsilon \mid \bullet T \mid [T]T \mid TA_1^{(1)}A_1^{(2)}A_2^{(1)}A_2^{(2)}$
	$T \to \epsilon \mid \bullet T \mid [T]$
	$\begin{pmatrix} A_1 \\ A_2 \end{pmatrix} \to \begin{pmatrix} A_1[T \\ ]TA_2 \end{pmatrix} \mid \begin{pmatrix} [T \\ ]T \end{pmatrix}$

LP+	$S \to \epsilon \mid \bullet T \mid [T]T \mid TA_1^{(1)}A_1^{(2)}A_2^{(1)}A_2^{(2)} \mid TA_1^{(1)}A_1^{(2)}A_2^{(1)}A_1^{(3)}A_2^{(2)}A_2^{(3)}$
	$\quad TA_1^{(1)}A_1^{(2)}A_1^{(3)}A_2^{(1)}A_2^{(2)}A_2^{(3)} \mid TA_1^{(1)}A_1^{(2)}A_2^{(1)}A_1^{(1)}A_2^{(4)}A_2^{(2)}A_2^{(3)}A_2^{(4)}$
	$T \to \epsilon \mid \bullet T \mid [T]$
	$\begin{pmatrix} A_1 \\ A_2 \end{pmatrix} \to \begin{pmatrix} A_1[T \\ ]TA_2 \end{pmatrix} \mid \begin{pmatrix} [T \\ ]T \end{pmatrix}$

DP	$S \to \epsilon \mid \bullet S \mid [S]S \mid A_1^{(1)}A_1^{(2)}A_2^{(1)}A_2^{(2)}$
	$\begin{pmatrix} A_1 \\ A_2 \end{pmatrix} \to \begin{pmatrix} A_1[S \\ ]SA_2 \end{pmatrix} \mid \begin{pmatrix} [S \\ ]S \end{pmatrix}$

RG	$S \to \epsilon \mid \bullet S \mid [S]S \mid A_1^{(1)}A_1^{(2)}A_2^{(1)}A_2^{(2)}$
	$\begin{pmatrix} A_1 \\ A_2 \end{pmatrix} \to \begin{pmatrix} A_1[S \\ ]A_2S \end{pmatrix} \mid \begin{pmatrix} [S \\ ]S \end{pmatrix}$

RN	$S \to \epsilon \mid \bullet S \mid [S]S \mid A_1^{(1)}A_1^{(2)}A_2^{(1)}A_2^{(2)} \mid A_1^{(1)}A_1^{(2)}A_2^{(1)}A_1^{(3)}A_2^{(2)}A_2^{(3)} \mid$
	$\quad A_1^{(1)}A_1^{(2)}A_1^{(3)}A_2^{(1)}A_2^{(2)}A_2^{(3)} \mid A_1^{(1)}A_1^{(2)}A_1^{(3)}A_2^{(1)}A_1^{(4)}A_2^{(2)}A_2^{(3)}A_2^{(4)}$
	$\begin{pmatrix} A_1 \\ A_2 \end{pmatrix} \to \begin{pmatrix} A_1[S \\ ]A_2S \end{pmatrix} \mid \begin{pmatrix} [S \\ ]S \end{pmatrix}$

AU	$S \to \epsilon \mid \bullet S \mid A_1A_2 \qquad\qquad \begin{pmatrix} A_1 \\ A_2 \end{pmatrix} \to \begin{pmatrix} M_1 \\ K_1M_2K_2 \end{pmatrix} \mid \begin{pmatrix} [S \\ ]S \end{pmatrix}$
	$\begin{pmatrix} M_1 \\ M_2 \end{pmatrix} \to \begin{pmatrix} M_1K_1^{(1)} \\ K_2^{(1)}K_1^{(2)}M_2K_2^{(2)} \end{pmatrix} \mid \begin{pmatrix} K_1 \\ K_2 \end{pmatrix} \qquad \begin{pmatrix} K_1 \\ K_2 \end{pmatrix} \to \begin{pmatrix} K_1[S \\ ]SK_2 \end{pmatrix} \mid \begin{pmatrix} [S \\ ]S \end{pmatrix}$

inclusive class of pseudoknots for which DP algorithms have become available so far [17]. On the other end of the spectrum, Lyngsø and Pedersen [11] considered non-recursive H-type pseudoknots. Below, we write T for the non-terminal describing pseudoknot-free structures. The LP+ model [4] includes also the four types of pseudoknots shown in Fig. 2. The model of Dirks & Pierce (DP) generalizes (LP) to include recursive H-type pseudoknots [5]. Reeder and Giegerich (RG) further restrict the appearance of unpaired bases in this setting [14]. While the two grammars look identical in Table 1, they differ in the implementation of the parsers for the terminals. This is due to a (recently remedied) limitation of the ADPfusion high-level parser that did not allow for interleaved non-terminals and terminals in the same "horizontal stack". The more efficient, original construction of RG in [14] is now possible, whereas the one in Table 1 disregards alternatives that do not fit into the RG scheme during parsing – which is semantically correct, but asymptotically suboptimal. "Simple pseudoknots" were defined by Akutsu and Uemura (AU) [1] as comprising two interleaving distinct sets of base pairs. These pairs create an interleaved stem within both groups. Base pairings are organized so that the first group's right bases and the second group's left bases are arbitrarily interleaved, while the other bases are all outside the interleaved area. Categorizing secondary structures by the topological genus, Reidys et al. [15] showed that there are exactly four types of pseudoknots with genus 1, the simplest of which is the H-type pseudoknot, see Fig. 2. The genus-1 structures are referred to as (RN) below.

In order to connect the pseudoknot grammars with Turner's standard energy model [27], we interpret $[S]$ and $[T]$ as a nonterminal B in the ViennaRNA recursions and employ the loop-decomposition of Eq. (2). Furthermore, we use the notation $A_1^{(i)}$ and $A_2^{(i)}$ for the components of two-dimensional non-terminals that have isomorphic productions (albeit possibly with different values in the evaluations algebras). For the latter we simply dropped the superscript $^{(i)}$ in Table 1.

In line with the simplified multiloop model, we consider a single parameter, namely a *pseudoknot initialization penalty*, Ψ, which is associated with all productions that introduce a 2-dimensional non-terminal on their left side. For all helical parts within pseudoknots, the standard stacking energies are used. Unpaired positions are assigned additive contributions corresponding to the multiloop model.

4   Computational Experiments

In order to evaluate the accuracy of pseudoknot prediction we used a subset of the RNAstrand database [2]. Due to the computational costs of the pseudoknot algorithms, which have asymptotic running times of $O(n^6)$, we restricted ourselves to entries with at most 70 nucleotides. This leaves 131 pseudoknot-free and 63 pseudoknotted target structures.

On the pseudoknot-free subset accuracy cannot exceed the accuracy on pseudoknot-free structures.[1] Very large values of Ψ, in fact, force the predictions to be pseudoknot-free. By construction, then, there is no difference between different grammars and the ViennaRNA-like baseline. On this data set, we achieve a limiting F1-measure of about 0.85 for $\Psi \geq 8$ kcal/mol. We note that this value is surprisingly large in comparisons with other benchmarks of RNA folding, probably due to the short sequences.

On the subset with pseudoknots, the performance does not depend very strongly on Ψ for moderate values, it decreases, however, for large values of $\Psi > 12$ kcal/mol as sensitivity decreases. This is expected, since excessive energy penalties for pseudoknots cause them to become markedly underpredicted.

Figure 3 summarizes the results. To give a balanced picture of performance and pseudoknotted and pseudoknot-free instances despite the difference in sample sizes, we averaged the performance measure for the two samples. As expected

---

[1] Since we use here an energy model that is slightly simplified in the evaluation of certain loop terms compared to the full model implemented in ViennaRNA, occasionally we predict structures that are closer to structure model in the STRAND database and thus accuracy may also be (slightly) better than the ViennaRNA predictions.

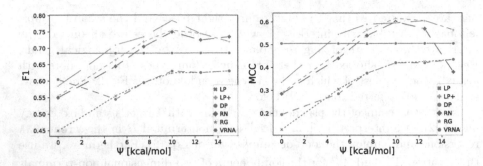

Fig. 3. Performance of five different MCFGs for pseudoknotted RNA structures in comparison with the pseudoknot-free background (VRNA). Both F1 and MCC show the best performance for $\Psi \approx 10$ kcal/mol and indicate a qualitative difference between the three grammars with recursive pseudoknots (LP+, RG, and RN) compared to DP and LP.

from analyzing the two subsets separately, we observe a performance peak for $\Psi \approx 10$ kcal/mol. Despite the short sequences in the test set we observe a superior performance of grammars that admit recursive pseudoknots.

5   Availability

This work is accompanied by git repositories. For readers who are interested in experimenting with pseudoknot grammars, we offer the "GenussFold" repository at https://github.com/choener/Prj-GenussFold. This project comes with all necessary dependencies and two options to experiment with and build pseudoknot grammars. It provides, via a `nix flake`, a complete development environment. In addition, if `nix` is not available, a more usual path via GHC Haskell and `cabal` is available. We refer to the readme in the project file on how to build the project. In addition, implementations for the different grammars are made available here: https://github.com/deggers/GenussFoldEnergyMin.

6   Concluding Remarks

Different grammars for the prediction of RNA structures with pseudoknots define vastly different search spaces. Variations of the grammar, therefore, include or exclude certain types of structures and thus in general will affect the predicted structures. While much effort has been expended to study and compare different implementations, no unifying framework was available in which all relevant pseudoknot model grammars are available together with a full fledged scoring system. As a consequence it has remained unclear to what extent differences in predictive power have to be attributed to the different scoring model, and to what extent the grammars themselves play an important role.

In this work we have begun a comprehensive study of the predictive power, advantages, and disadvantages due to the choice of grammar. So far, our study has been constrained to a subset of six grammars, including the pseudoknot-free RNA folding grammars from the ViennaRNA package [9]. Furthermore, we had to restrict ourselves to the set of sequences that can be folded by *all* grammars within predefined resource limits in order to accurately compare the quality of predictions. In order to minimize the influence of differences in scoring models we used here the initialization energy Ψ for a pseudoknot as the single free parameter and otherwise treated pseudoknots like multiloops. Prediction performance as a function of Ψ suggests a plausible value of about 10 kcal/mol for the optimal choice of this parameter. Interestingly, this value matches well with regression-based multiloop initialization terms, see [28] for an overview of multiloop energy models. We also observed that the two grammars LP and DP that do not admit recursive pseudoknots are outperformed by the three grammars that include recursive pseudoknots. Given the short size of the benchmarking targets this is surprising and deserves a closer examination.

This first study exposes several avenues for further exploration. When we began this study, we noted that certain production rules did not fit immediately into our framework. We chose to rewrite grammars to fit into the framework, while keeping their semantics intact. Since then, progress in ADPfusion ameliorates these shortcomings. A forthcoming more detailed study hence will encompass the full range of pseudoknot grammars. Recent improvements in the parsing and compiler fusion system further optimizes the resulting program code, enabling a systematic benchmark on significantly longer input sequences and thus more difficult instances.

Inspection of the grammars in Table 1 shows that the grammars are composed of many common rules or parts of rules. This suggests to make systematic use of another feature of ADP, namely the capability to compose grammars by additions, subtractions, and multiplications [22]. This type of construction will provide a guarantee that subsets of grammars that are supposed to be equal, will indeed generate the same structures, while at the same time reduce the complexity of the algorithms themselves. This approach will also simplify the exploration of more sophisticated energy models for pseudoknots, which in the simplest case distinguish the initialization terms for different knot types as suggested e.g. in [15].

Finally, the ability to automatically generate outside grammars opens up the possibility of calculating ensemble quantities and provides an important building block for parameter training extensions. The latter are required as grid based searches, as we performed for the pseudoknot initialization penalty, do not scale beyond two or three independent parameters.

References

1. Akutsu, T.: Dynamic programming algorithms for RNA secondary structure prediction with pseudoknots. Discr. Appl. Math. **104**, 45–62 (2000). https://doi.org/10.1016/S0166-218X(00)00186-4

2. Andronescu, M., Bereg, V., Hoos, H.H., Condon, A.: RNA STRAND: the RNA secondary structure and statistical analysis database. BMC Bioinf. **9**, 340 (2008). https://doi.org/10.1186/1471-2105-9-340
3. Brierley, I., Pennell, S., Gilbert, R.J.: Viral RNA pseudoknots: versatile motifs in gene expression and replication. Nat. Rev. Microbiol. **5**, 598–610 (2007). https://doi.org/10.1038/nrmicro1704
4. Condon, A., Davy, B., Rastegari, B., Zhao, S., Tarrant, F.: Classifying RNA pseudoknotted structures. Theor. Comp. Sci. **320**, 35–50 (2004). https://doi.org/10.1016/j.tcs.2004.03.042
5. Dirks, R.M., Pierce, N.A.: A partition function algorithm for nucleic acid secondary structure including pseudoknots. J. Comput. Chem. **24**, 1664–1677 (2003). https://doi.org/10.1002/jcc.10296
6. Giegerich, R., Meyer, C.: Algebraic dynamic programming. In: Kirchner, H., Ringeissen, C. (eds.) Algebraic Methodology And Software Technology (AMAST 2002), vol. 2422, pp. 243–257. Springer, Berlin (2002). https://doi.org/10.5555/646061.676145
7. Giegerich, R., Meyer, C., Steffen, P.: A discipline of dynamic programming over sequence data. Sci. Comput. Prog. **51**, 215–263 (2004). https://doi.org/10.1016/j.scico.2003.12.005
8. Giegerich, R., Touzet, H.: Modeling dynamic programming problems over sequences and trees with inverse coupled rewrite systems. Algorithms **7**, 62–144 (2014). https://doi.org/10.3390/a7010062
9. Lorenz, R., et al.: ViennaRNA package 2.0. Alg. Mol. Biol. **6**, 26 (2011). https://doi.org/10.1186/1748-7188-6-26
10. Lyngsø, R.B., Pedersen, C.N.: RNA pseudoknot prediction in energy-based models. J. Comp. Biol. **7**, 409–427 (2000). https://doi.org/10.1089/106652700750050862
11. Lyngsø, R.B., Pedersen, C.N.: Pseudoknots in RNA secondary structures. In: Shamir, R., Miyano, S., Sorin, I. (eds.) RECOMB 2000: Proceedings of the Fourth Annual International Conference on Computational Molecular Biology, pp. 201–209. ACM, New York (2000). https://doi.org/10.1145/332306.332551
12. Nebel, M.E., Weinberg, F.: Algebraic and combinatorial properties of common RNA pseudoknot classes with applications. J. Comp. Biol. **19**, 1134–1150 (2012). https://doi.org/10.1089/cmb.2011.0094
13. Ponty, Y., Saule, C.: A combinatorial framework for designing (pseudoknotted) RNA algorithms. In: Przytycka, T.M., Sagot, M.-F. (eds.) WABI 2011. LNCS, vol. 6833, pp. 250–269. Springer, Heidelberg (2011). https://doi.org/10.1007/978-3-642-23038-7_22
14. Reeder, J., Giegerich, R.: Design, implementation and evaluation of a practical pseudoknot folding algorithm based on thermodynamics. BMC Bioinf. **5**, 104 (2004). https://doi.org/10.1186/1471-2105-5-104
15. Reidys, C.M., Huang, F.W.D., Andersen, J.E., Penner, R.C., Stadler, P.F., Nebel, M.E.: Topology and prediction of RNA pseudoknots. Bioinformatics **27**, 1076–1085 (2011). https://doi.org/10.1093/bioinformatics/btr090, addendum. In: Bioinformatics 28:300 (2012)
16. Riechert, M., Höner zu Siederdissen, C., Stadler, P.F. Algebraic dynamic programming for multiple context-free grammars. Theor. Comp. Sci. **639**, 91–109 (2016). https://doi.org/10.1016/j.tcs.2016.05.032
17. Rivas, E., Eddy, S.R.: A dynamic programming algorithm for RNA structure prediction including pseudoknots. J. Mol. Biol. **285**, 2053–2068 (1999). https://doi.org/10.1006/jmbi.1998.2436

18. Rivas, E., Lang, R., Eddy, S.R.: A range of complex probabilistic models for RNA secondary structure prediction that include the nearest neighbor model and more. RNA **18**, 193–212 (2012). https://doi.org/10.1261/rna.030049.111
19. Seki, H., Matsumura, T., Fujii, M., Kasami, T.: On multiple context free grammars. Theor. Comp. Sci. **88**, 191–229 (1991). https://doi.org/10.1016/0304-3975(91)90374-B
20. Sheikh, S., Backofen, R., Ponty, Y.: Impact of the energy model on the complexity of RNA folding with pseudoknots. In: Kärkkäinen, J., Stoye, J. (eds.) CPM 2012. LNCS, vol. 7354, pp. 321–333. Springer, Heidelberg (2012). https://doi.org/10.1007/978-3-642-31265-6_26
21. Höner zu Siederdissen, C.: Sneaking around concatMap: efficient combinators for dynamic programming. In: Proceedings of the 17th ACM SIGPLAN International Conference on Functional Programming, ICFP 2012, pp. 215–226. ACM, New York (2012). https://doi.org/10.1145/2364527.2364559
22. Höner zu Siederdissen, C., Hofacker, I.L., Stadler, P.F.: Product grammars for alignment and folding. IEEE/ACM Trans. Comp. Biol. Bioinf. **12**, 507–519 (2014). https://doi.org/10.1109/TCBB.2014.2326155
23. Höner zu Siederdissen, C., Prohaska, S.J., Stadler, P.F.: Algebraic dynamic programming over general data structures. BMC Bioinf. **16**, S2 (2015). https://doi.org/10.1186/1471-2105-16-S19-S2
24. Staple, D.W., Butcher, S.E.: Pseudoknots: RNA structures with diverse functions. PLoS Comp. Biol. **3**, e213 (2005). https://doi.org/10.1371/journal.pbio.0030213
25. Steffen, P., Giegerich, R.: Versatile and declarative dynamic programming using pair algebras. BMC Bioinf. **6**, 224 (2005). https://doi.org/10.1186/1471-2105-6-224
26. Taufer, M., et al.: PseudoBase++: an extension of PseudoBase for easy searching, formatting, and visualization of pseudoknots. Nucl. Acids Res. **37**, D127–D135 (2009). https://doi.org/10.1093/nar/gkn806
27. Turner, D.H., Mathews, D.H.: NNDB: the nearest neighbor parameter database for predicting stability of nucleic acid secondary structure. Nucl. Acids Res. **38**, D280–D282 (2010). https://doi.org/10.1093/nar/gkp892
28. Ward, M., Datta, A., Wise, M., Mathews, D.H.: Advanced multi-loop algorithms for RNA secondary structure prediction reveal that the simplest model is best. Nucl. Acids Res. **45**, 8541–8550 (2017). https://doi.org/10.1093/nar/gkx512

Comparison of Machine Learning Pipelines for Gene Expression Matrices

Mateus Devino(iD), Kele Belloze(iD), and Eduardo Bezerra(✉)(iD)

Federal Center for Technological Education of Rio de Janeiro (CEFET/RJ),
Rio de Janeiro, Brazil
`mateus.pereira@aluno.cefet-rj.br`, {`kele.belloze,ebezerra`}`@cefet-rj.br`

Abstract. Advances in genetic sequencing technologies have enabled the understanding of the course of diseases in a manner like never before. These technologies produce a data structure called an expression matrix, which contains gene expression values taken under certain sampling conditions. In this paper, we present preliminary work on comparing the application of different machine learning pipelines to an expression matrix. As a case study, we consider a dataset from the Gene Expression Omnibus containing gene expression levels (obtained through scRNA-seq) in the context of Breast Cancer disease. We present a generalized processing pipeline instantiation and discuss the corresponding results.

Keywords: Machine learning · scRNA-seq · Breast cancer

1 Introduction

Cancer is a set of genetic diseases characterized by an uncontrollable growth of cells. Normal cells die when replication errors trespass a certain threshold. Cancer cells seem to avoid control mechanisms and continue to replicate uncontrollably, even when replication errors accumulate [7]. Cancer is the leading cause of death worldwide, with about ten million deaths in 2020 [10]. Any part of the body can develop cancer. Breast cancer was the fifth type that most killed in 2020, with nearly 685 thousand deaths [10]. It was also the type of cancer with the highest number of registered new cases in the same year.

Due to the heterogeneity of cancer, there is no one-size-fits-all treatment. Even the same type of cancer may have several subtypes that differ considerably. On top of that, each person is unique, and individual genetic variation may influence the outcome of treatments. Both characteristics have proven personalized approaches more successful than generic treatments [5].

With the advances in genetic sequencing, it is possible to understand the course of diseases like never before. RNA sequencing [4] techniques have been used over the past decades, resulting in humongous amounts of data about the human body and many genetic diseases.

The traditional method of RNA sequencing (RNA-seq) analyzes bulks of cells, resulting in each bulk's gene expression being represented by the mean of

N. M. Scherer and R. C. de Melo-Minardi (Eds.): BSB 2022, LNBI 13523, pp. 32–37, 2022.
https://doi.org/10.1007/978-3-031-21175-1_4

expression levels in each cell in bulk. This type of analysis is useful in most cases. However, due to cancer's heterogeneity, means can mask important information about individual cell's expression levels [8]. As computational resources become more accessible, Single Cell RNA sequencing (scRNA-seq) becomes possible.

In contrast with the traditional RNA-seq, scRNA-seq captures gene expression levels for each cell individually. While genetic sequencing is not a perfect process, the technology has become more accessible over the years, resulting in growing amounts of data being generated every year. As more data is generated, Machine Learning techniques have been used to advance the current understanding of cancer [3].

Machine Learning (ML) is a subarea of Artificial Intelligence (AI) in which computers are not explicitly programmed to perform a task. Instead, a model learns to perform such a task by looking at examples. The more quality data is available, the better results yielded by ML models tend to be. In this paper, we report ongoing research whose objective is to analyze how different scRNA-seq data pipelines perform in identifying potential therapeutic targets for treating breast cancer.

This paper is divided into four sections, including this introduction. Section 2 describes the methodology applied in this study and the minimal background needed to understand the research. Section 3 presents the results achieved so far. Finally, Sect. 4 summarizes what was presented in the paper and lists future steps for this ongoing research.

2   Methodology

This research aims to analyze how different scRNA-seq pipelines support drug discovery for Breast Cancer. The pipeline created for our experiments currently consists of four steps: (i) gene filtering, (ii) normalization, (iii) dimensionality reduction, and (iv) cell clustering. Figure 1 illustrate these steps, which will be further discussed in this section.

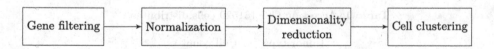

Fig. 1. Single Cell RNA-sequencing analysis pipeline high-level steps.

Gene Filtering. Gene expression data is noisy. For that reason, as a first step in the pipeline, we define a filtering activity in which lowly expressed genes are discarded from the analysis. Several filtering alternatives can be used to instantiate this step, e.g., mean, variance, and log-fold change.

Normalization. Normalization is a set of data transformation techniques to prepare the data to be ingested by an ML model. Many ML algorithms expect the input data to adhere to specific constraints. Normalization techniques must be chosen based on the algorithm to be used, as well as the problem to be solved.

Dimensionality Reduction. Dimensionality reduction, as the name suggests, consists of applying techniques to reduce the number of features (genes) of each sample (cells). This technique is essencial in the scRNA-seq analysis as each cell has expression levels for hundreds of thousands genes.

Cell Clustering. Clustering is an unsupervised machine learning technique that aims to group similar samples. Our scRNA-seq case study applies clustering techniques to group tumor and control cells in distinct clusters.

3   Experiments and Results

The experiments were conducted on an Windows Subsystem for Linux (WSL) Arch Linux distribution, with an Intel i9-10900k CPU with 10 cores (20 threads) and 64GB RAM. The pipeline was implemented using Scikit Learn [9]. We first describe the dataset used (Sect. 3.1), followed by the pipeline instantiation (Sect. 3.2) and main results of the experiment (Sect. 3.3).

3.1   Dataset

The data utilized in our research is available in the Geo Expression Omnibus database (GEO), accession GSE161529[1]. It consists of approximately 430 thousand samples (cells) collected from 69 scRNA-seq profiles performed in 52 patients. Different types of tumors can be found on the dataset, such as TNBC, ER+, HER+, BRCA1 TNBCs and lymph-node metastases. The distribution of cells for each class, grouped by breast cancer subtype, can be found in Table 1.

Table 1. Accession GSE161529 cells distribution.

Tumor type	Cell count
ER+	162,160
Normal	122,800
HER2+	48,010
TNBC BRCA1	42,767
BRCA1	26,272
TNBC	22,616
PR+	3,406

[1] https://www.ncbi.nlm.nih.gov/geo/query/acc.cgi?acc=GSE161529.

The following sections describe the experiments executed. All cells of the previously described GSE161529 dataset were used. A pipeline containing the following steps was analyzed: (i) mean expression level for gene filtering, (ii) maximum absolute scaling for normalization, (iii) incremental PCA for dimensionality reduction, and (iv) K-Means for clustering. Although differential cluster analysis is a step planned to be performed by the pipeline, it has not been performed at this time.

3.2   Pipeline Instantiation

Gene Filtering → Mean Expression Level. In the gene filtering step, we removed all the genes with zero mean expression level. As a result, from the total of 33538 genes in the original matrix, 2522 genes were discarded.

Normalization → Maximum Absolute Scaling. In preparation for PCA, it is important to scale the gene expression values, so all genes have the same weight. Due to the dataset can only be loaded in memory in a sparse format, maximum absolute scaling was applied to the data not to disrupt its sparsity. The operation performed by the maximum absolute scaling is $y = \frac{x}{max(x)}$. In this equation, x denotes the gene expression level for a given cell, $max(x)$ denotes the highest expression level for the same gene in all cells, and y denotes the scaled value, which is a value between -1 and 1. As gene expression levels cannot be negative, our case study's values are always non-negative. As a result of the maximum absolute scaling, each gene in the expression matrix has its expression levels at the same magnitude.

Dimensionality Reduction → PCA Principal Component Analysis (PCA) technique identifies which components (linear combinations of features) most explain the variance of a given dataset. Its output is a list of features ranked, from highest to lowest, by the amount of variance explained.

A limitation of the traditional PCA is that the whole dataset must be loaded in memory. As the dataset contains about 430 thousand cells, with expression levels for about 30 thousand genes, the whole dataset could only be loaded in memory as a sparse matrix. The traditional PCA does not work with sparse matrices, so Incremental PCA [1] was used. We use 295 components in our experiments.

Clustering → K-Means. K-Means starts by defining K centroids, data points in the same space as the given dataset. Then, each sample is assigned to the cluster whose centroid is the closest. After all samples have been assigned, the mean of each cluster is calculated. The means are elected as the new centroids, and each sample is reassigned to a cluster based on them. The process repeats until there is no significant change between iterations [6]. In the current pipeline instantiation, we use $K = 7$, since we have 6 types of cancerous cells plus normal cells.

There are different techniques to initialize centroids, which are stochastic. This initialization influences the final result [2]. Also, notice that centroids are not necessarily a sample in the dataset but a data point in the same dimensional space.

3.3 Results

The pipeline described above was trained with all cells in GSE161529. The hyperparameters utilized for PCA were 295 components, and 1024 batch size. For K-Means, the K value chosen was 7. Figure 2 presents the pipeline steps and the hyperparameters used.

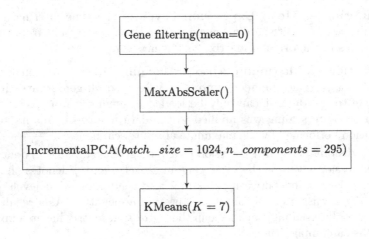

Fig. 2. Pipeline instantiation and hyperparameters.

The silhouette coefficient was applied as a metric to evaluate the quality of the clustering yielded by the pipeline. The silhouette coefficient is calculated by using $s(o) = \frac{b(o)-a(o)}{max\{a(o),b(o)\}}$. In this equation, o denotes a cell, $a(o)$ denotes the average distance between o and all cells in the same cluster as o, and $b(o)$ denotes the the distance between o and all cells not in the same cluster as 0.

The silhouette coefficient results in a value between -1 and 1, and our pipeline achieved a 0.6274 upon clustering the cells in the dataset. The code to execute the experiments is available at GitHub[2].

4   Final Remarks

Due to the heterogeneous nature of cancer and individual genetic variation, scRNA-seq analysis can be an important tool towards understanding cancer. This paper described ongoing research aimed at analyzing different scRNA-seq pipelines applied to breast cancer patients data. The current ongoing step in this

---

[2] https://github.com/MLRG-CEFET-RJ/brca-scrna-seq/tree/BSB2022.

research is implementing the differential cluster analysis. The expected outcome of this step is to highlight over-expressed genes in cancer cells, which would be potential therapeutic targets.

Once the differential cluster analysis is complete, diverse techniques could be applied in each step of the existing pipeline. This variation would generate new pipelines and results, which could be compared against the existing one. Also, quality metrics should be applied to filter bad quality cells before training a new pipeline instances.

References

1. Artac, M., Jogan, M., Leonardis, A.: Incremental pca for on-line visual learning and recognition. In: 2002 International Conference on Pattern Recognition, vol. 3, pp. 781–784 (2002)
2. Arthur, D., Vassilvitskii, S.: k-means++: The advantages of careful seeding. Technical report, Stanford (2006)
3. Asada, K., et al.: Single-cell analysis using machine learning techniques and its application to medical research. Biomedicines 9(11), 1513 (2021)
4. Conesa, A.: A survey of best practices for rna-seq data analysis. Genom. Biol. 17(1), 13 (2016)
5. Ding, S., Chen, X., Shen, K.: Single-cell rna sequencing in breast cancer: understanding tumor heterogeneity and paving roads to individualized therapy. Cancer Commun. 40(8), 329–344 (2020)
6. Han, J., Pei, J., Tong, H.: Data Mining: Concepts and Techniques. Morgan kaufmann, Burlington (2022)
7. Hanahan, D.: Hallmarks of cancer: new dimensions. Cancer Disc. 12(1), 31–46 (2022)
8. Li, X., Wang, C.Y.: From bulk, single-cell to spatial rna sequencing. Int. J. Oral Sci. 13(1), 1–6 (2021)
9. Pedregosa, F., et al.: Scikit-learn: machine learning in python. J. Mach. Learn. Res. 12, 2825–2830 (2011)
10. Sung, H.: Global cancer statistics 2020: globocan estimates of incidence and mortality worldwide for 36 cancers in 185 countries. CA Cancer J. Clin. 71(3), 209–249 (2021)

Evaluating Machine Learning Models for Essential Protein Identification

Jessica da Silva Costa(iD), Jorge Gabriel Rodrigues(iD), and Kele Belloze$^{(\boxtimes)}$(iD)

Federal Center for Technological Education of Rio de Janeiro (CEFET/RJ),
Rio de Janeiro, Brazil
jessica.costa.1@aluno.cefet-rj.br , kele.belloze@cefet-rj.br

Abstract. Drug development is often a complex and time-consuming process. Especially in the initial phase, selecting a target for drug development can take many years. Essential genes and proteins are biological entities responsible for the biological processes of survival and reproduction of organisms. Studies indicate that essential genes tend to have higher expression and encode proteins that engage in more protein-protein interactions. All these characteristics make essential proteins potential drug targets. Thus, this work proposes using protein-protein interaction-based features to train and evaluate machine learning algorithms to identify essential proteins. Experiments with the organism *Saccharomyces cerevisiae* indicate that the application of the Random Forest algorithm and balancing techniques obtained better recall values.

Keywords: Machine learning · Protein-protein interaction · Essential protein

1 Introduction

The development of a new drug, from the original idea to the launch of a final product, is a complex process that can take 12 to 15 years [2]. The entire drug discovery process during clinical trials takes much time because there are multiple testing phases. Especially the first phase can be very costly to build a supporting body of evidence before selecting a target for an expensive drug discovery program [2,12]. On the other hand, there is currently a vast collection of publicly available biological databases [7] that greatly assist in *in silico* research for targeted drug development. In this context, essential genes and proteins are potential drug targets.

Genes considered necessary for the survival or reproduction of an organism are classified as essential genes [1,3]. Essential genes have wide applications in the pharmaceutical field, and their encoded proteins also have important roles in several vital functions of organisms [3,5]. Essential genes also tend to be more expressed and encode proteins that engage in more protein-protein interactions (PPI) [11]. Given this situation, detecting essentiality in genes or proteins is one of the first tasks in discovering drug targets.

N. M. Scherer and R. C. de Melo-Minardi (Eds.): BSB 2022, LNBI 13523, pp. 38–43, 2022.
https://doi.org/10.1007/978-3-031-21175-1_5

Several work in the literature present computational methods (or *in silico* searches) related to identifying essentiality in genes and proteins. Some work use protein-protein interaction (PPI) networks, gene expression, homology, sequence features, and machine learning in predictive algorithms to predict essentiality in genes and proteins [3,5,10,16]. In general, the works combine several types of inputs and not a specific one to make predictions.

Machine learning (ML) deals with algorithms that learn through training data from a problem. It is an area that has wide application in bioinformatics [13]. Specifically, in gene and protein essentiality prediction, some works use PPI data in Machine Learning algorithms [8,9]. In this view, this on going work proposes using protein-protein interaction-based features to train and evaluate machine learning algorithms to identify essential proteins. We present an instantiation with *Saccharomyces cerevisiae* data, an eukaryotic well-studied organism.

The article is organized into four sections, including this introduction. Section 2 presents the proposed research methodology and its minimal background. Section 3 presents the results achieved, and Sect. 4 summarizes the final remarks and the next steps for this work.

2 Methodology

This work aims to evaluate machine learning models to identify essential proteins. This work's methodology proposes using centrality measures and clustering in graphs as model features. The following activities are conducted this work: (i) Data Preparation and (ii) Data Training and Prediction.

2.1 Data Preparation

PPI networks are complex, and one possible way to characterize them is through measures, for example, clustering or centrality. Centrality is one of the fundamental principles of network analysis. It measures how "central" a node is in the network. For this work, the input features of machine learning models are based on the calculation of the measures: *Degree Centrality*, *Eigenvector Centrality*, *Betweenness Centrality*, *Closeness Centrality* and *Clustering*. The calculations considered the graph formed by the PPI network of the experimental organism.

The PPI data utilized in our research is available in the STRING database [6]. The measurements were calculated using the NetworkX library [15]. After the calculations, we label the data with the aid of an integrated essentiality dataset built based on the DEG database [14] to label training and test data. This integrated dataset (available at Github[1]) consists of 6,394 proteins of the experimental organism, 1,100 essential, and 5,294 non-essential.

---

[1] https://github.com/Jessicalta/deg-data-essentiality.

2.2 Data Training and Prediction

Five algorithms were selected for applying ML models: K-Nearest Neighbors (KNN), Support Vector Machine (SVM), Random Forest, Decision Tree and Logistic Regression. They were selected based on related work. The hyperparameter search was performed using the grid search method and stratified cross-validation, given that the classes (essential and non-essential) are unbalanced. The stratified division of training and testing data classes was performed with 70% of the database for training and 30% for testing. In addition to using the original data, Oversampling and Undersampling techniques were applied to balance the classes. Random Undersampling and SMOTE (Synthetic Minority Oversampling Technique) were applied for Undersampling and Oversampling, respectively.

Algorithm implementations from the Scikit-Learn library were used. Specifically for Decision Trees, the CART implementation is available [17]. Two metrics are used for selecting the best hyperparameters: accuracy and AUC-ROC. Accuracy measures the correctness of a classifier. The AUC-ROC is a measure that summarizes the probability curve of the rate of true positives concerning the rate of false positives [4]. The best hyperparameter results are selected for the prediction phase with test data. High accuracy in the model can be influenced by the majority class, in this case, non-essential. Therefore, in this work, we adopted precision and recall to evaluate the model in identifying essentiality in proteins.

3   Results

This section presents the results found according to the proposed methodology. Figure 1 shows the data characteristics of the PPI network's calculated measures. In *Closeness Centrality* the essential proteins are concentrated between 0.4 and 0.6. As for the clustering index, the values range from 0.1 to 0.8, while the non-essential are more distributed between 0 and 1. In the *Betweenness Centrality*, the presence of outliers is noticed. Figure 2 presents the correlation between the features of the model through a heatmap graph. Although the classes are unbalanced, eigenvector, centrality, and closeness measures have the highest positive correlations for essentiality. These results indicate that the three measures may have greater weight in the prediction.

Table 1 presents the accuracy metrics and AUC ROC achieved in each model in the training data. A group of hyperparameters with the best result in training data for each algorithm and balancing technique is selected for the predictions. In total, 30 groups of hyperparameters were found, from which 15 groups were selected for prediction in the test group. Considering the application of Oversampling or Undersampling techniques for balancing classes, it is important to highlight that the best hyperparameters were achieved with AUC ROC in the cross-validation. Table 2 shows precision and recall results of the selected models.

The evaluation shows the original data have high recall for the non-essentiality class and low recall for essentiality. The results also show that the

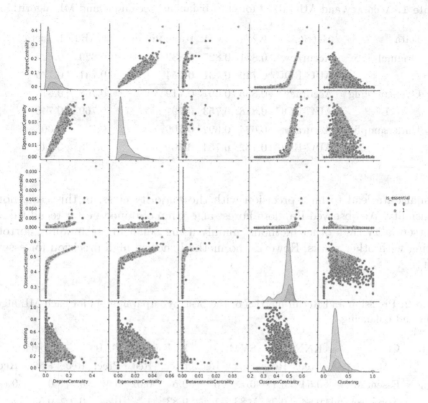

Fig. 1. Distribution of data from the calculated measures of the PPI network

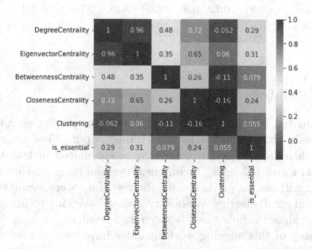

Fig. 2. Correlation of data from the calculated measures of the PPI network

Table 1. Accuracy and AUC ROC for each balancing technique and ML algorithm

Data	Metric	KNN	SVM	Random forest	CART	Logistic
Original data	Accuracy	0.831	0.827	0.832	0.829	0.827
	AUC-ROC	0.719	0.731	0.754	0.741	0.735
Oversampling	Accuracy	0.835	0.697	0.719	0.708	0.67
	AUC-ROC	0.878	0.753	0.791	0.769	0.746
Undersampling	Accuracy	0.674	0.692	0.706	0.7	0.665
	AUC-ROC	0.732	0.754	0.76	0.738	0.743

original data lead to high precision with the majority class, in this case, non-essentiality. We observed Random Forest algorithm obtained better recall values for essentiality than other methods, mainly using balancing techniques, corroborating with other works. However, no method showed high precision for essentiality.

Table 2. Precision and Recall metrics of the best hyperparameters for each ML algorithm and balancing

Data	Class	KNN		SVM		R. Forest		CART		Logistic	
		Prec.	Rec.	Prec.	Rec.	Prec.	Rec.	Prec.	Rec.	Prec.	Rec.
Orig.	Essential	0.54	0.1	0	0	0.5	0.02	0.46	0.07	0	0
	Non essential	0.84	0.98	0.83	1	0.83	1	0.84	0.9	0.83	1
Over	Essential	0.26	0.43	0.29	0.69	0.28	0.78	0.27	0.72	0.3	0.65
	Non essential	0.86	0.74	0.91	0.65	0.93	0.58	0.91	0.59	0.9	0.68
Under	Essential	0.27	0.72	0.3	0.64	0.28	0.76	0.29	0.7	0.3	0.65
	Non essential	0.91	0.6	0.9	0.69	0.92	0.59	0.91	0.64	0.9	0.69

4   Final Remarks

This work evaluated Machine Learning algorithms to predict essential proteins using PPI data's characteristics as algorithms' input. Five algorithms were selected, and their best hyperparameters were calculated in training set with the original data and balancing techniques: oversampling and undersampling. The chosen algorithms for prediction in the test data were evaluated based on the precision and recall metrics. We observed that balancing techniques with the Random Forest algorithm obtained better recall results.

The next step of this ongoing work includes improvements related to input data of the classifier algorithm, adding protein sequence features. After that, new experiments will be conducted applying neural network algorithms. At the end of this research, the expected outcome is applying the machine learning predictive model in organisms with unknown essential proteins and supporting the first phase of the drug development process.

References

1. Zhang, Z., Ren, Q.: Why are essential genes essential?-the essentiality of Saccharomyces genes. Microbial Cell **2**(8), 280 (2015)
2. Hughes, J.P., et al.: Principles of early drug discovery. Brit. J. Pharmacol. **162**(6), 1239–1249 (2011)
3. Peng, C., et al.: A comprehensive overview of online resources to identify and predict bacterial essential genes. Front. Microbiol. **8**, 2331 (2017)
4. Huang, J., Ling, C.X.: Using AUC and accuracy in evaluating learning algorithms. IEEE Trans. Knowl. Data Eng. **17**(3), 299–310 (2005). https://doi.org/10.1109/TKDE.2005.50
5. Belloze, K., et al.: A review of artificial neural networks for the prediction of essential proteins. Netw. Syst. Biol., 45–68 (2020)
6. Szklarczyk, D., et al.: The STRING database in 2021: customizable protein-protein networks, and functional characterization of user-uploaded gene/measurement sets. Nucleic Acids Res. **49**(D1), D605–D612 (2021). https://doi.org/10.1093/nar/gkaa1074
7. Rigden, D.J., Fernández, X.M.: The 2022 nucleic acids research database issue and the online molecular biology database collection. Nucleic Acids Res. **50**(D1), D1–D10 (2022)
8. Azhagesan, K., et al.: Network-based features enable prediction of essential genes across diverse organisms. PloS one **13**(12), e0208722 (2018). https://doi.org/10.1371/journal.pone.0208722
9. Zhang, J., et al.: NetEPD: a network-based essential protein discovery platform. Tsinghua Sci. Technol. **25**(4), 542–552 (2020)
10. Garcia, F.P., Guedes, G.P., Belloze, K.T.: Identifying *Schistosoma mansoni* essential protein candidates based on machine learning. In: Kowada, L., de Oliveira, D. (eds.) BSB 2019. LNCS, vol. 11347, pp. 123–128. Springer, Cham (2020). https://doi.org/10.1007/978-3-030-46417-2_12
11. Wang, T., et al.: Identification and characterization of essential genes in the human genome. Science **350**(6264), 1096–1101 (2015)
12. Biswas, R., et al.: Drug discovery and drug identification using AI. In: 2020 Indo-Taiwan 2nd International Conference on Computing, Analytics and Networks (Indo-Taiwan ICAN). IEEE (2020)
13. Srinivasa, K.G., Siddesh, G.M., Manisekhar, S.R. (eds.): Statistical Modelling and Machine Learning Principles for Bioinformatics Techniques, Tools, and Applications. AIS, Springer, Singapore (2020). https://doi.org/10.1007/978-981-15-2445-5
14. Luo, H., et al.: DEG 15, an update of the database of essential genes that includes built-in analysis tools. Nucleic Acids Res. **49**(D1), D677–D686 (2021)
15. Hagberg, A., Pieter S., Chult, D.S.: Exploring network structure, dynamics, and function using NetworkX. No. LA-UR-08-05495; LA-UR-08-5495. Los Alamos National Lab. (LANL), Los Alamos, NM (United States) (2008)
16. Aromolaran, O., et al.: Essential gene prediction in Drosophila melanogaster using machine learning approaches based on sequence and functional features. Comput. Struct. Biotechnol. J. **18**, 612–621 (2020)
17. Pedregosa, F., et al.: Scikit-learn: machine learning in python. J. Mach. Learn. Res. **12**, 2825–2830 (2011)

Study on the Complexity of Omics Data: An Analysis for Cancer Survival Prediction

Carlos Daniel Andrade[1][iD], Thomas Fontanari[1,2][iD],
and Mariana Recamonde-Mendoza[1,2(✉)][iD]

[1] Institute of Informatics, Universidade Federal do Rio Grande do Sul,
Porto Alegre, RS, Brazil
{tvfontanari,mrmendoza}@inf.ufrgs.br
[2] Bioinformatics Core, Hospital de Clínicas de Porto Alegre, Porto Alegre, RS, Brazil

Abstract. The use of machine learning approaches in studying cancer through omics datasets has been an important research tool since the advent of high-throughput technologies. However, these datasets present an intrinsic data complexity that may hinder model development despite their information richness. This work, therefore, aims to study the characteristics of different omics data commonly employed for clinical predictive analysis using a broad set of data complexity measures tailored for imbalanced domains. We focus on the task of cancer survival prediction in eight tumor types based on four types of omics data (*i.e.,* copy number variation, gene expression, microRNA expression, and DNA methylation) and the combination among them (*i.e.,* multi-omics approach). We found that F1-MaxDr, F3_partial, F4_partial, and N3_partial could be used as predictors of performance in this scenario. Furthermore, our experiments suggested that the studied omics data types are strongly correlated in terms of data complexity, including the multi-omics approach. All eight cancer types appeared to be highly correlated with each other, except for Adrenocortical Carcinoma (ACC), which showed a significantly lower complexity than the others in the analyzed data.

Keywords: Complexity measures · Omics data · Multi-omics · Cancer

1 Introduction

Cancer diagnosis and clinical prognosis are issues of profound importance in current medicine. For patients diagnosed with cancer, in particular, the prognosis is a matter of special interest since accurately predicting the chance of survival can help doctors make better treatment decisions [6]. Given that cancer is known to

This study was financed in part by the Coordenação de Aperfeiçoamento de Pessoal de Nível Superior - Brasil (CAPES) - Finance Code 001, and by grants from the Fundação de Amparo á Pesquisa do Estado do Rio Grande do Sul (FAPERGS) [21/2551-0002052-0] and Conselho Nacional de Desenvolvimento Científico e Tecnológico (CNPq) [308075/2021-8].

N. M. Scherer and R. C. de Melo-Minardi (Eds.): BSB 2022, LNBI 13523, pp. 44–55, 2022.
https://doi.org/10.1007/978-3-031-21175-1_6

be related to an individual's genetic characteristics, the omics datasets constitute a powerful tool for further investigating the disease and improving its prevention, diagnosis, treatment, and prognosis. Numerous advances for predicting diagnosis or patient survival from omics data have been powered by concurrent progress in machine learning (ML) techniques [15]. Omics datasets, however, present intrinsic characteristics that hinder the construction of reliable ML models.

Two well-known characteristics of omics datasets are their high dimensionality and class imbalance [3,11,14]. However, other forms of characterizing the complexity of datasets in a broader scope have been developed in the last years [1,2,5,8]. These *complexity measures* assess characteristics such as class overlap, data sparsity, and the complexity of the decision boundary, and may be applied to obtain scientific insights into the domain or even conduct data-driven choices regarding pre-processing and classification techniques.

Previous works analyzed complexity measures for transcriptomics (Sect. 2). However, there has been no exploration of the intrinsic complexity of other omics data types, nor a comparison between cancer types in terms of their omics-based complexity. Furthermore, complexity measures have been recently adapted to better assess data complexity in class-imbalanced scenarios [1,2] - a particularity that was not considered by previous analyses on gene expression data.

Therefore, the goal of this work is to conduct an initial investigation of the intrinsic characteristics of omics datasets across multiple types of omics and cancers. We focus our study in the task of predicting 3-year overall survival in patients with cancer, analyzing complexity measures in eight cancer types and five distinct omics datasets (*i.e.,* mRNA expression, microRNA expression, copy number variation (CNV), DNA methylation, and multi-omics). We compare and discuss the data complexity measures concerning the minority class (*i.e.,* non-survivors), assessing their utility as predictors of classification performance. Moreover, we make comparisons among types of cancers and omics data to assess if any differences or similarities exist regarding data intrinsic characteristics.

2   Related Works

The difficulty of classification problems is often linked to data complexity, which arises due to data intrinsic characteristics. As means to assess aspects that influence data complexity, Ho and Basu [5] proposed an initial set of twelve measures to evaluate geometrical characteristics of the class distributions, including the overlap of individual features, separability of classes and geometry, topology, and density of manifolds. This original set was further reviewed and extended by Lorena *et al.* [8], who also proposed standardizing metrics in bounded intervals (*i.e.,* [0, 1]) where higher values indicate greater complexity, thus making them more easily comparable and interpretable. Later, Barella *et al.* [1] adapted complexity measures to consider the class imbalance found in many datasets. In what follows, we review the main studies that explored data complexity measures, adjusted or not for data imbalance, in the domain of omics data.

Okun and Priisalu [11] presented one of the first works involving complexity analysis of omics data. Authors identified an association between data complexity and the performance of k-nearest neighbors (k-NN) models applied to gene expression datasets for binary classification in cancer diagnosis. Lorena *et al.* [9] studied how data scarcity affects the performance of classifiers in the task of gene marker selection from microarray datasets. Data scarcity is a complexity measure given by the ratio between dimensionality and the number of samples [5].

De Souto *et al.* [14] conducted a more comprehensive analysis of the difficulty of cancer diagnosis prediction using microarray data, adopting more complexity measures and evaluating their correlation with the error rates of a Support Vector Machine (SVM) classifier. This investigation was further expanded to address the issue of class imbalance, feature correlation, and the impact of feature selection [7]. Lorena *et al.* [7] found that data sparsity and class imbalance are important factors influencing classification performance and that dimensionality reduction by feature selection tends to decrease the impact of these characteristics.

The correlation between complexity measures and classification performance based on microarray data was also identified in other studies [3,10]. Morán-Fernández *et al.* [10] showed that measures related to class overlap were positively associated with classification accuracy, while measures reflecting characteristics of class boundary and dispersion among samples were particularly helpful in predicting the performance of k-NN models. Moreover, their results corroborate the fact that feature selection may reduce data complexity. Finally, Sánchez and Garcia [13] sought to study the relationship between the curse of dimensionality and intrinsic data characteristics, such as class overlapping and class separability, using gene expression data as a typical example of high-dimensional data.

None of these works, however, analyzed other types of omics data besides gene expression profiles (*i.e.*, transcriptomics), despite their growing availability and use as features in clinical prediction models [12]. In addition, the aforementioned works were developed before the changes proposed by Barela *et al.* [1] to adapt the metrics to class imbalance scenarios - a recurrent characteristic in biomedical and omics data. In this sense, our work differs from the previous ones in that it seeks to study other types of omics data besides gene expression data, including multi-omics data, and compare complexity measures across types of cancers and omics. Furthermore, we use complexity metrics adapted to class imbalanced data to obtain more robust results for the analysis of omics data.

3   Data Complexity Measures

Although several efforts were made to standardize and review data complexity measures [2,5,8], this work follows the definitions provided by Barella *et al.* [2], which adapted the original measures for class imbalanced scenarios. The original definitions were shown to be inappropriate for imbalanced datasets because the majority class dominates measures calculation, hiding the behavior of other classes that may actually be more interesting, depending on the domain. The modified complexity measures compute the original measures per class, yielding a set of C values for each measure, where C is the number of classes (hence the use of the suffix partial below) [2]. This adaptation allows an analysis focused on the class of interest (usually the minority class). Finally, we note that the complexity measures output a value between 0 and 1, such that higher values indicate higher data complexity and, thus, harder classification tasks.

Feature-Based Measures. Four different feature-based measures were used, which aim at describing how informative the features are. *Maximum Fisher's Discriminant Ratio* (F1-MaxDr) calculates how much overlap exists between classes for each predictive feature by using the Fisher discriminant ratio. *Volume Overlapping Region* (F2_partial) is based on the hyper-volume of the overlapping region between classes. Finally, *Individual Feature Efficiency* (F3_partial) measures the complexity in terms of the individual efficiency of each feature to separate the classes, and *Collective Feature Efficiency* (F4_partial) follows a similar idea as F3_partial but is based on computing feature efficiencies after the other features have been used to separate the data.

Neighborhood-Based Measures. The neighborhood-based measures attempt to analyze the boundary between classes. In this work, we considered five measures in this category. The *Fraction of Points on the Class Boundary* (N1_partial) assesses the fraction of instances that are close to the class boundary using a minimum spanning tree. *Ratio of Average Intra/Inter Class NN Distance* (N2_partial) compares intraclass and interclass dispersion by computing their ratio. The dispersion formulas are based on the distance between nearest neighbors (NN) of similar and different classes. *Leave-one-out Error Rate of the NN Classifier* (N3_partial) is based on the error rate of the 1-NN algorithm computed using leave-one-out cross-validation. *Non-linearity of a 1-NN Classifier* (N4_partial) is similar to N3_partial but is performed on a synthesized dataset constructed by interpolating instances from the same class. Finally, the *Fraction of Maximum Covering Spheres* (T1_partial) consists in growing hyperspheres around each instance until it touches a hypersphere of a different class. Hyperspheres contained within larger ones are eliminated, and the ratio of remaining hyperspheres and instances is used to estimate the metric value.

Linearity-Based Measures. The linearity measures aims to assess whether the task is linearly separable. We considered three linearity-based measures. *Minimized Sum of Error Distance of a Linear Classifier* (L1_partial) constructs a linear classifier (such as linear SVM, as used in this work) and calculates the average distance between the wrongly classified instances and the separating hyperplanes for each class – greater distances imply in higher complexity. *Training Error of a Linear Classifier* (L2_partial) is also based on a linear classifier, but instead of using the distances, it simply computes the error rate of the linear classifier for each class. Finally, *Non-linearity of the Linear Classifier* (L3_partial) uses a similar technique as N4_partial. It creates an interpolated dataset and uses them to evaluate the error rate of the model.

4 Methodology

Our goal is to study the intrinsic characteristics of omics data for predicting cancer prognosis through data complexity measures adapted for class imbalance.

Specifically, we addressed the 3-year survival prediction task for eight types of cancer, each containing four different omics datasets and a multi-omics dataset. In this section, we present our methodology for data collection and analysis. All analyses were performed using the R programming language.

4.1   Collection and Preparation of Omics Data

We used the omics datasets shared by Duan *et al.* [4]. Authors collected four types of omics data from The Cancer Genome Atlas (TCGA) for several tumors: copy number variation (CNV) at the genome level, messenger RNA (mRNA) at the transcriptome level, and DNA methylation (methy) and micro-RNA (miRNA) at the epigenome level. The choice for these data types was due to their frequent use in studies developing diagnosis and prognosis models for cancer. Raw data was pre-processed by authors to remove batch effects, filter our features with more than 20% of missing values, impute remaining missing values with k-NN algorithm, and normalize values using z-scores. Moreover, we chose the *significant* version of the datasets [4], which has a reduced number of feature selected based on statistical analysis.

We analyzed eight cancer types: Adrenocortical Carcinoma (ACC), Invasive Breast Carcinoma (BRCA), Colon Adenocarcinoma (COAD), Renal Papillary Cell Carcinoma (KIRP), Clear Cell Renal Carcinoma (KIRC), Hepatocellular Carcinoma (LIHC), Lung Adenocarcinoma (LUAD), and Squamous Cell Lung carcinoma (LUSC). For a given cancer, all omics data types share the same instances, that is, each individual has CNV, mRNA, Methy, and miRNA data. Furthermore, we created a multi-omics dataset for each type of cancer by simply concatenating the array values for the four types of omics per individual. The motivation for generating this dataset is based on the increasing attention given to the integration of multi-omics data as inputs to inform precision medicine-based decision-making [12]. For more details about the datasets and the pre-processing steps, we refer reader to the work by Duan *et al.* [4].

The collected datasets, however, did not contain clinical information for the patients. Therefore, we obtained the clinical variables from FireBrowse[1], which contains data generated by the TCGA project. In particular, we obtained information on vital status and survival time for each tumor sample in the datasets and synthesized this information into a new binary target attribute indicating whether the patient had survived three years after the tumor diagnosis (3-year survival). We kept only the instances for which this information was available. Table 1 summarizes the information about each dataset used in this work.

4.2   Extraction of Data Complexity Measures

We extracted the data complexity measures using the R packages ECoL[2] and ImbCoL[3]. For each of the 40 datasets summarized in Table 1 (*i.e.,* five omics datasets

---

[1] http://firebrowse.org/.

[2] https://github.com/lpfgarcia/ECoL.

[3] https://github.com/victorhb/ImbCoL.

Table 1. Number of samples per class for each cancer type and number of features for each omics type. Note that the individuals (*i.e.*, instances) are shared across all omics types given a cancer type, such that the number of instances are the same for all omics.

| | 3-Year Survival | | | No. features | | | | |
	Yes	No	Total	mRNA	miRNA	Methy	CNV	Multi-omics
ACC	60	17	77	2000	200	2000	524	4274
BRCA	707	49	756	2000	200	2000	1974	6174
COAD	241	48	489	2000	200	2000	1449	5649
KIRC	239	68	307	2000	200	2000	2102	6302
KIRP	246	25	271	2000	200	2000	1023	5223
LIHC	263	101	364	2000	200	2000	2050	6250
LUAD	328	116	444	2000	200	2000	3446	7646
LUSC	246	108	354	2000	200	2000	3074	7274

and eight cancer types), a total of 12 data complexity measures were calculated, namely: feature-based metrics F1-MaxDr, F2_partial, F3_partial, and F4_partial; neighbourhood-based metrics N1_partial, N2_partial, N3_partial, N4_partial, and T1_partial; and the linearity-based metrics L1_partial, L2_partial, and L3_partial. With the exception of F1-MaxDr that has no imbalanced version, we analyzed the metric value obtained for the class of interest, which in the case of the target attribute 3-year survival, refers to non-survivors (*i.e.*, negative) class.

4.3   Training and Evaluation of the Predictive Models

Model training and evaluation was performed using the R package caret. Two algorithms were selected to train predictive models for 3-year survival: Naive Bayes (NB) and generalized linear models (GLM) fitted using a boosting approach (the glmboost method in R). The choice of these algorithms was due to the simplicity in terms of hyperparameters to be adjusted. The models were trained for each type of omics data, including the multi-omics dataset, and for each type of cancer using 5-fold cross validation repeated 10 times.

We evaluated the models considering as the class of interest (commonly called the *positive* class) the No class, which represents patients who did not survive for more than three years after tumor diagnosis. The focus on this class is based on the importance of identifying tumor cases with a worse prognosis as soon as possible, allowing for a more careful medical follow-up in order to try to prevent such adverse outcome. We analyzed recall, precision, and F1-score for each combination of cancer type and omics dataset.

5   Results

5.1   Complexity Measures and Classification Performance

Figure 1 shows the dataset complexities according to measures F1-MaxDr, F3_partial, and N3_partial (F4_partial was omitted due to large similarity with F3_partial).

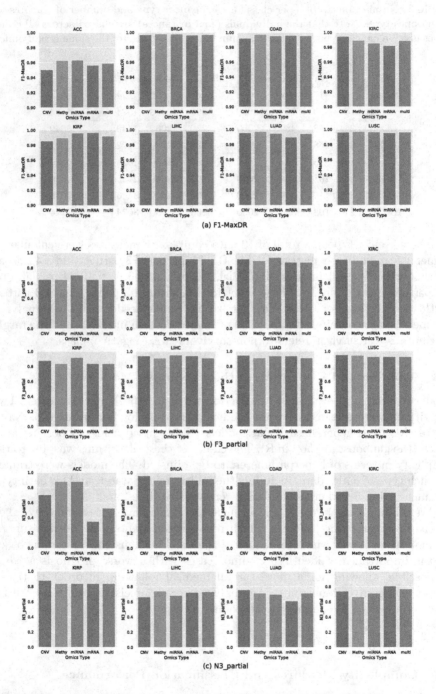

Fig. 1. Complexity of each cancer type and omics data for F1_MaxDR, F3_partial, and N3_partial.

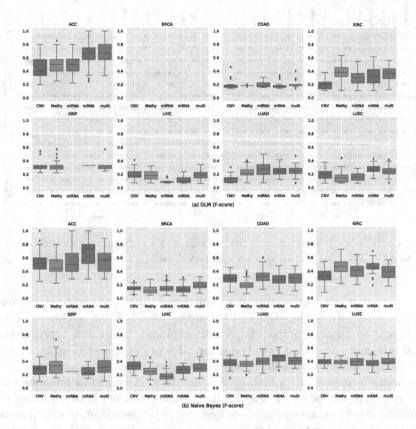

Fig. 2. F1-scores for each dataset and omics data using GLM and Naive Bayes.

The complexity measures showed in Fig. 1 had the highest correlations with classification performance metrics and thus, due to space constraints, were selected for our analyses[4]. In general, few differences were observed among omics datasets for the same type of cancer, implying that the difficulty in determining cancer prognostic based on data of this nature is similar among all types of omics for a given cancer type. Complexity measures showed some degree of variation for ACC, KIRC, and KIRP when analyzing F1-MaxDr; KIRP when analyzing F3_partial; and ACC, COAD, KIRC, LIHC, LUAD, and LUSC when analyzing N3_partial. Thus, among the complexity measures showed in Fig. 1, N3_partial had the largest variation across different omics datasets for each cancer type. The F1-scores obtained using GLM and NB for each cancer and omics data type are shown in Fig. 2 and the correlations between complexity measures and classification performance are presented in Fig. 3.

---

[4] The raw results of our experiments can be found in the project Github repository: https://github.com/carlosdanielandrade/complexity-of-omics-data-in-cancer.

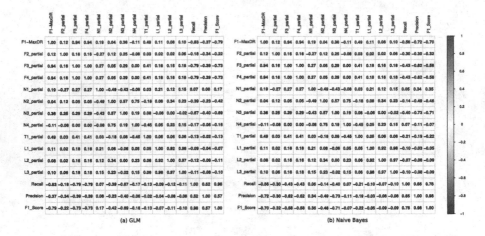

Fig. 3. Correlations between complexity measures and performance metrics using (a) GLM and (b) NB. The correlations are computed using all cancers and omics types.

We note that due to the large class imbalance, GLM suffered from low recall for BRCA, thus setting the value of F1-score to zero. It is interesting to observe that there is no standard in the relative performance of omics data types, as there is no single omics data that stands out for all types of cancer (Fig. 2). In many cases, we have observed that the multi-omics dataset does not increase predictive performance as expected - a finding that has been discussed in previous works [4]. Moreover, as can be seen in Fig. 3, almost all correlations between complexity measures and performance metrics are negative, as expected, since more complex datasets (values closer to 1) should indicate a more difficult task (performance closer to 0).

Furthermore, the two models (GLM and NB) showed similar trends for both complexity and performance measures. As aforementioned, F1-MaxDr, F3_partial, F4_partial, and N3_partial showed the strongest correlations with the performance of the classification models. In some situations, these measures can work as predictors of performance level. In this sense, a lower complexity may indicate a trend towards higher performance. Finally, it is also interesting to note that N1_partial has a positive correlation with performance. For the GLM, these values are not very significant, but become more relevant in the case of NB.

5.2 Comparison Among Omics and Cancer Types

In this Section, we consider how omics data types and cancer types compare in terms of their complexity measures. Specifically, we use F1-MaxDr, F3_partial, and N3_partial, since those were found to be the most correlated measures with performance. Figure 4 shows swarm plots of these complexity measures. The left side of the figure shows the measures grouped by cancer type, such that each point corresponds to the complexity of an omics data for a specific cancer group. Similarly, the right side of the figure groups the measures by omics data type.

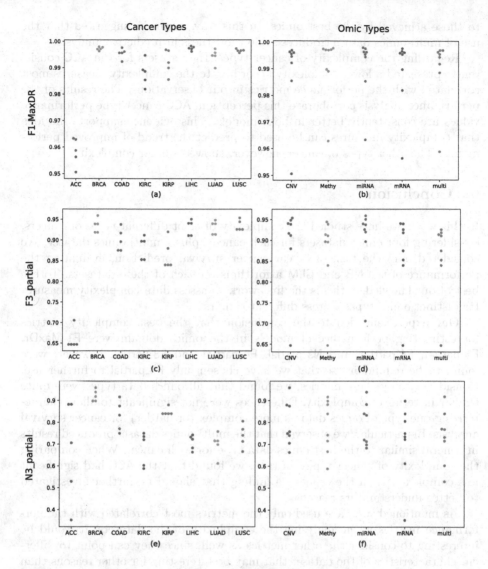

Fig. 4. Swarm plots showing complexity measures grouped by cancer type in (a), (c), and (e) and omics data type in (b), (d), and (f).

It is possible to observe that the behavior of the different omics types is quite similar for the measures considered, corroborating the idea of similar complexity among distinct omics datasets. More interestingly, we note that the use of the multi-omics approach did not result in less complexity overall. However, the multi-omics dataset never has the result below all other omics. Additionally, often within a cancer type, multi-omics data have results quite close to or equal to those achieved by the best omics. In this way, it can be suggested that the use of multi-omics data introduces some robustness in predictive analysis.

Regarding the complexity of cancer types, the datasets for the ACC consistently presented a lower complexity according to the complexity measures most correlated with the performance metrics in our observations. The results of the performance analysis corroborate this perception. ACC cancer type performance values are consistently better in both models. This evidence supports the idea that complexity measures can be used to predict the trend of improved performance. The other types of cancer, however, showed similar complexity.

6   Conclusion

In this work, we have studied the complexity of eight different types of cancers, considering four omics datasets for each cancer, plus a multi-omics dataset. We considered only the task of 3-year cancer survival prediction, evaluating the performance of the NB and GLM algorithms on each of these datasets. To the best of our knowledge, this is the first work to assess data complexity measures for distinct omics types, across different tumors.

Our experiments led to the conclusion that the best complexity metrics for estimating performance of models in the omics domain were F1-MaxDr, F3_partial, F4_partial, and N3_partial. F4_partial and F3_partial, however, were found to be redundant, so that we have chosen only F3_partial to further use. Considering only these metrics, we found that all omics data types were quite similar in terms of complexity. Differences were not significant to allow characterizing one type of omics data as more complex (or harder) for cancer survival analysis. In particular, we observed that the multi-omics dataset produced results in general similar to the best omics data type for each cancer. When comparing the complexity of cancer types, in turn, we found that the ACC had significant less complexity than the others - a finding that should be further investigated to better understand its reasons.

As mentioned, we have used only the metrics most correlated with the performance metrics when analyzing the complexity of the datasets. It would be interesting to consider the other metrics as well, since they can point to different characteristics of the dataset that may be interesting for other reasons than performance estimation. Furthermore, we have considered only two classifiers, and one could naturally ask whether the found metrics would remain the most correlated if different classifiers were considered. Finally, we chose survival prediction as our task of interest given its importance in the context of cancer and omics-based prediction, but one could also consider other tasks. We are currently working on extensions of this work to take these considerations into account.

References

1. Barella, V.H., Garcia, L.P., de Souto, M.C., Lorena, A.C., de Carvalho, A.C.: Assessing the data complexity of imbalanced datasets. Inf. Sci. **553**, 83–109 (2021)

2. Barella, V.H., Garcia, L.P., de Souto, M.P., Lorena, A.C., de Carvalho, A.: Data complexity measures for imbalanced classification tasks. In: 2018 International Joint Conference on Neural Networks (IJCNN), pp. 1–8. IEEE (2018)
3. Bolón-Canedo, V., Moran-Fernandez, L., Alonso-Betanzos, A.: An insight on complexity measures and classification in microarray data. In: 2015 International Joint Conference on Neural Networks (IJCNN), pp. 1–8. IEEE (2015)
4. Duan, R., et al.: Evaluation and comparison of multi-omics data integration methods for cancer subtyping. PLOS Comput. Biol. **17**(8), 1–33 (2021)
5. Ho, T.K., Basu, M.: Complexity measures of supervised classification problems. IEEE Trans. Pattern Anal. Mach. Intell. **24**(3), 289–300 (2002)
6. Li, J., et al.: Predicting breast cancer 5-year survival using machine learning: a systematic review. PLOS ONE **16**(4), 1–23 (2021)
7. Lorena, A.C., Costa, I.G., Spolaôr, N., De Souto, M.C.: Analysis of complexity indices for classification problems: cancer gene expression data. Neurocomputing **75**(1), 33–42 (2012)
8. Lorena, A.C., Garcia, L.P., Lehmann, J., Souto, M.C., Ho, T.K.: How complex is your classification problem? a survey on measuring classification complexity. ACM Comput. Surv. **52**(5), 1–34 (2019)
9. Lorena, A.C., Spolaor, N., Costa, I.G., Souto, M.C.P.: On the complexity of gene marker selection. In: 2010 Eleventh Brazilian Symposium on Neural Networks, pp. 85–90 (2010)
10. Morán-Fernández, L., Bolón-Canedo, V., Alonso-Betanzos, A.: Can classification performance be predicted by complexity measures? a study using microarray data. Knowl. Inf. Syst. **51**(3), 1067–1090 (2017)
11. Okun, O., Priisalu, H.: Dataset complexity in gene expression based cancer classification using ensembles of k-nearest neighbors. Artif. Intell. Med. **45**(2–3), 151–162 (2009)
12. Olivier, M., Asmis, R., Hawkins, G.A., Howard, T.D., Cox, L.A.: The need for multi-omics biomarker signatures in precision medicine. Int. J. Molec. Sci. **20**(19), 4781 (2019)
13. Sánchez, J.S., García, V.: Addressing the links between dimensionality and data characteristics in gene-expression microarrays. In: Proceedings of the International Conference on Learning and Optimization Algorithms: Theory and Applications, pp. 1–6 (2018)
14. de Souto, M.C.P., Lorena, A.C., Spolaôr, N., Costa, I.G.: Complexity measures of supervised classifications tasks: a case study for cancer gene expression data. In: The 2010 International Joint Conference on Neural Networks (IJCNN), pp. 1–7 (2010)
15. Zhao, D., et al.: Pan-cancer survival classification with clinicopathological and targeted gene expression features. Cancer Inf. **20**, 11769351211035137 (2021). pMID: 34376966

Identifying Large Scale Conformational Changes in Proteins Through Distance Maps and Convolutional Networks

Lucas Moraes dos Santos[1,2]([✉]) [iD] and Raquel C. de Melo Minardi[1,2] [iD]

[1] Department of Computer Science, Federal University of Minas Gerais,
Belo Horizonte, Minas Gerais, Brazil
`lucasmds@ufmg.br`, `raquelcm@dcc.ufmg.br`
[2] Department of Biochemistry and Immunology, Federal University of Minas Gerais,
Belo Horizonte, Minas Gerais, Brazil

Abstract. Conformational changes in protein structures are strongly correlated with functional changes. Some conformational modifications may be easily noticeable, others are more subtle. In this work, we model the problem of protein conformation classification through its representation as images that illustrate the interatomic distance matrices. We aim then to discover if a convolutional neural network would be able to identify these conformational changes only from the distance patterns in these maps. Hence, this work presents the development of a model based on convolutional neural networks, capable of identifying large scale conformational changes in proteins. As a case study, we used the S protein from SARS-CoV-2, a protein known for its function in the infection of human cells through a conformational change to binding to the human cell receptor. Initially, we intend to identify large-scale conformations, such as states where the S protein trimers are closer together (closed) or further away (open). The proposed classifier achieved a satisfactory performance after cross validation, reaching an average accuracy in validation of 90.58%, with an error of 22.31%. The model was also able to successfully distinguish both classes (open and closed states for S protein) achieving a precision of 84.32% and a recall of 89%. In the test, the accuracy of the model reached 71.79%, with an error rate of 28.2%. Precision and recall reached 68.18% and 78.94%, respectively. For future work, we want to evaluate the ability of such model to identify even more subtle conformational changes, as well as those caused by point mutations that occur in virus variants.

Keywords: Conformational changes · Distance maps · Convolutional networks

1 Introduction

The emergence of new algorithms with potential application in computational biology has significantly contributed to the advancement of research related

© The Author(s), under exclusive license to Springer Nature Switzerland AG 2022
N. M. Scherer and R. C. de Melo-Minardi (Eds.): BSB 2022, LNBI 13523, pp. 56–67, 2022.
https://doi.org/10.1007/978-3-031-21175-1_7

to human health, the understanding of biological systems and biotechnological advances [31].

Furthermore, it is observed that the areas of biological and health sciences have produced a growing volume of data due to technological advances in omics sciences (genomics/metagenomics, proteomics, transcriptomics, metabolomics, interatomic, etc.) [8]. As an example, the *Protein Data Bank* (PDB) [7,15] currently provides 202.228 biological macromolecular structures, 7.416 only in the first 7 months of the year.

Mathematical and computational approaches applied to the study of protein features are quite useful in structural bioinformatics. An interest and robust model of protein structures is distance matrix [22]. A distance matrix $(\mathbf{d}_{ij})$ is obtained by calculating the distance between the ith and the jth C^{α} (α–carbons) of the amino acid residues belonging to the protein chains. Applications in structural bioinformatics involving the alignment of protein structure or to infer protein-protein interactions used distance matrices [14,33]. From them, it is possible to generate distance maps that is, a 2D visual representation of these matrices [23].

More recently, the use of artificial intelligence (AI) has stood out, mainly of deep neural networks, in solving problems related to structural computational biology [3,4,20]. Recent applications of Deep Learning (DL) in bioinformatics have been used to gain insights from data, which has been emphasized both in academia and industry. Artificial intelligence has been used for over a decade both in modeling biological data and catalyzing new discoveries [30] in the field of molecular biology.

In this sense, recent approaches have emerged, such as AlphaFold2, a DL-based artificial intelligence program developed by Google's DeepMind that builds theoretical models of protein structures [2,20]. In addition to Alphafold, recent works involving protein structure prediction have used distance maps in the development of models based on DL [3,4,34]. These models were able to recognize complex patterns during the training process with large datasets.

Recent works involving the classification or structural prediction of proteins have performed these analyzes based on distance maps [2–4,20,34].

Among the widely used DL algorithms there are the Convolutional Networks (ConvNet's) [24], characterized by a specialized type of neural network for data processing that has a topology similar to a grid. The architecture of a CNN is analogous to the pattern of connectivity of neurons in the human brain. Being inspired by the organization of the visual cortex, it stands out as an example of neuroscientific principles that influence deep learning [16].

Convolution is a linear operator that, given two functions, results in a third which measures the sum of the product of these functions along the region implied by their overlap as a function of the displacement between them [16]. Let f and g be continuous functions at time t, and u be a displacement, the convolution s can be represented mathematically as

$$s(t) = (f * g)(t) = \int_{-\infty}^{\infty} f(u)g(t - u)du. \tag{1}$$

In digital image processing applications, the input is commonly represented by a multidimensional array of data (an image, for example), and the kernel as a multidimensional array of parameters that are adapted by the learning algorithm. These data structures are referred to as *tensors* [16]. Let $x_{i,j}$ be a two-dimensional image and w be a kernel of dimensions l and k, the value of the convolution at any point (i, j) in the image can be described as

$$w \star x_{i,j} = \sum_l \sum_k w_{l,k} x_{i-l,j-k}. \tag{2}$$

This class of neural networks has shown great potential in applications involving pattern recognition in images [16], being used recently in conformational analysis, structure prediction, protein classification, etc [26]. The basic structure of CNNs consists basically in two parts: feature learning (convolution layers, non-linear layers and pooling layers) and classification (flatten layer and fully connected layer) [26].

Given this, and due to the current pandemic scenario, this work proposes the development of AI-based models capable of identifying conformational changes in SARS-CoV-2 (coronavirus) proteins through distance maps and convolutional neural networks. In particular, we evaluate if the CNNs can identify Spike protein (S) in open and close conformation through distances maps.

The COVID-19 (Coronavirus Disease) is a respiratory disease caused by the severe acute respiratory syndrome coronavirus 2 (SARS-CoV-2 - betacoronavirus belonging to the coronavirus family - *coronaviridae*, recently named as *new coronavirus*). The virus has zoonotic origin and the first known case of the disease dates back to December 2019 in Wuhan, China [25,37,39].

On January, 2020, the World Health Organization (WHO) classified the outbreak as an International Public Health Emergency and, on March, 2020, as a pandemic. Worldwide, more than 562 million cases of the disease were recorded, and a total of about 6.37 million deaths were due to complications of COVID-19 [12]. There are numerous computational challenges that permeate the models and algorithms that will be developed to support computational biology developments aiming at a better understanding of SARS-CoV-2 genome, proteins and variants, as well as to aid in drug and vaccine development for this or other emerging pathogens.

The main contributions of this work are: i) modeling the problem of classification of protein conformational changes as an image classification problem ii) design and implementation of a convolutional neural network capable of discriminating between protein conformations iii) the application of the proposed methodology to a case study involving the S protein of SARS-CoV-2, responsible for the entry of the pathogen into the human cell.

This work is organized as follows: in the next section we present the work methodology, starting with the collection and processing of protein data followed by the creation of distance maps. Next, we present the topology and parameterization decisions of the proposed network and the experimental design. This section is followed by the results and discussions and, finally, the conclusions and perspectives of the work.

2   Methodology

The methodological approach developed in this work can be summarized in the following steps: data collect and alignment for labeled the structures, generation of distance maps, implementation of the model based on deep learning (ConvNet) and model evaluation (Fig. 1).

2.1   Dataset

2.1.1   Data Collection in the PDB and Its Processing

We collect files in *.pdb* format (*Protein Data Bank* - PDB) [7] containing the S protein of SARS-CoV-2 (betacoronavirus belonging to the coronavirus family - *coronaviridae*, recently named as new coronavirus) [25,37,39], from the public database   RCSB PDB - Research Collaboratory for Structural Bioinformatics PDB. These files are composed of protein sequences and atomic three-dimensional coordinates, molecular compounds (small molecules and peptides) as well as their interactions with target proteins.

To generate the training labels, one of the proteins was used as a model (in this case the SARS-CoV-2 Spike PDB ID 6VYB [35]). We manually identified the *open* and *closed* chains for this file. Each of the other files were processed and protein chains were extracted, and superposed with model chains.

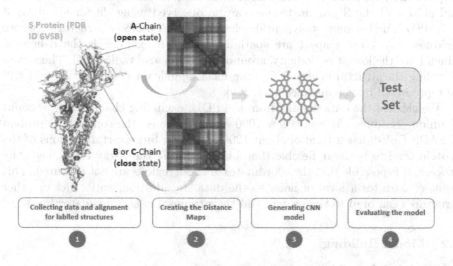

Fig. 1. The methodological approach developed can be summarized in the following steps: data collect and alignment for labeled the structures, generation of distance maps, implementation of the ConvNet and model evaluation.

TM-align is an algorithm for obtaining the best superposition between a pair of proteins through a rotation matrix built with the TM-Score and dynamic programming [38]. The TM-align source code was downloaded and the C++ version of the program was installed, with two metrics extracted to label the protein chains: TM-Score and RMSD. The TM-score can have a value between (0.1], where 1 indicates a perfect superposition between two structures [38]. The RMSD (Root Mean Square Deviation) is an average measure of the distances, or deviations, of a set of atoms. This metric is used to assess how similar (or dissimilar) are two proteins. RMSD is used to compute the deviation between the equivalent atom sets [23]. We labeled the protein chains according to the most similar model protein superposed. If a chain was closer conformationally to the open model, them we labeled it as *open*. On the contrary, we labeled it as *close*.

2.1.2   Generation of Distance Maps

We generated *.png* image files (Portable Network Graphics), with visual representation of the *Distance Map* of the S protein chains. These images are produced by extracting the coordinates (x, y, z) of the residues C_α (alpha carbons) and calculating the Euclidean distance between all pairs residues [22,23]. Once the intra-chain distances are calculated, we generate an matrix of distances of dimensions 309×309. An example distance map is shown in the Fig. 2 for both protein states (*open* and *closed*).

We sampled one for every 3 C_α. This was necessary to adapt the distance maps to the ConvNet input. Some large-scale conformations, such as the opening and closing of the S protein trimer, can be observed using thresholds above 3 Å RMSD. Furthermore, polypeptide chains contain periodic structures, where residues 3 or 4 units apart are spatially close. An example is the α-helices, which have the lowest periodicity, around 3,6 amino acid radicals [6]. Thus, even sampling the structure every 3 C_α, the most significant conformation will still be represented by the distance maps [19].

We cleaned the data coming from the PDB, excluding chains whose amount of unique residues $C\alpha$ was below 1000 alpha carbons (the complete S protein of SARS-CoV-2 has a total of about 1200 residues). Since certain regions of the protein tend to be more flexible than others, during the X-ray crystallography process, it is possible that the coordinates of these regions are not captured. This event constitutes a form of noise to the data. In addition, antibodies or other structures can bind to the protein and their removal is also mandatory.

2.2   Model Building

2.2.1   Model Architecture

The algorithms for obtaining distance maps and artificial intelligence models were implemented using the Python programming language, machine learning libraries and consolidated neural networks libraries such as TensorFlow and Keras [1,9]. Since the distance map is the result of transforming the 2D distance matrix into images, it was decided to develop models based on convolutional networks [2–4,20,34].

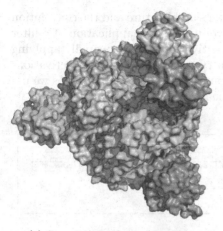

(a) S protein in the open state.

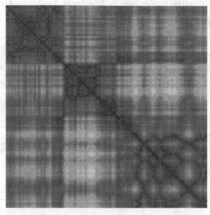

(b) Distance map corresponding to the chain A of the S protein.

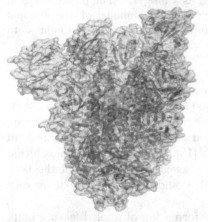

(c) S protein in the close state. Here, it is possible to observe the arrangement of the protein chains as well.

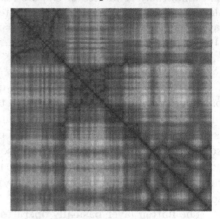

(d) Distance map corresponding to the chain B of the S protein.

Fig. 2. (a) 3D representation of the S protein (PDB ID 6VSB) in the open state. A-chain (orange) is further away from B-chain and C-chain (green); (b) Distance map between pairs of α–carbons residues of the A-chain of protein S. The axes correspond to the residues used to obtain the distance matrix; (c) 3D representation of the S protein (PDB ID 6VSB) in the closed state. It is possible to observe the conformational variation since A-chain is closer to B-chain and C-chain (closed state); (d) Distance map between pairs of α–carbons residues of the B-chain of protein S. (Color figure online)

The architecture of the model (shown in the Fig. 3) is based on the approach known as *representation learning* [5], in which a system automatically extracts the characteristics needed for classification/detection from the raw data. The convolution layer is the first layer in the model's architecture and the convolution operation is performed in this layer. A convolution is the application of a filter over an input (image) that can result in an activation [16]. Repeatedly applying the same filter (known as kernel) to an entry results in a map of activations called feature map. To increase the non-linear properties of feature maps we use an activation function, ReLU (Rectified Linear Unit) [16].

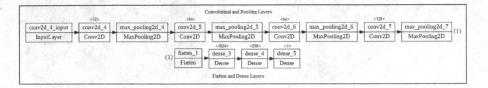

Fig. 3. ConvNet architecture developed. The number of filters (convolutional layers) and units (fully connected layers) are represented as <*integer*>. The pooled maps in the output (1) serve as input to the flatten layer (1). As the number of convolutional layers increases, it is possible to extract more complex features from the input data [16].

To reduce the variance for small changes in the image and the amount of parameters trained by the network (a technique commonly known as down sampling), pooling layers were used. Among existing pooling operations, max-pooling is often widely used. Max-pooling is a operation where the maximum value from kernel is extracted from the area it convolves [16]. Combined with convolution and pooling layers, dropout layer is used to prevent certain parts of the neural network from having too much responsibility and, consequently, being very sensitive to small changes [28].

The flatten layer basically operates a transformation of pooled feature map, changing its format to an array. The, dense layer, also called fully-connected layer, refers to the layer whose inside neurons connect to every neuron in the preceding layer. The activation function defined for this layer was sigmoid [13].

2.2.2   Model Parametrization

The convolution operation is applied to the input images, and to the pooled maps, using a 3×3 *kernel*, with a stride equal to 1. The Activation maps generated from the convolution layer serve as input to the pooling layer, where max-pooling was applied with a 2×2 pool array, with a stride equal to pooling size [9].

The defined batch size is 32. This number defines the number of samples that are propagated through the network. A larger batch size requires more memory, so it is common to find batch sizes equal to 32, or 64 [20,27]. Structures called tensors serve as input data for training the models. A tensor is composed of the input shape (image dimensions - height and width - and number of channels - RGB is equal to 3) and batch size [10].

The images were normalized with each pixel converted to a range of $[0-1]$, since neural networks tend to operate better with values in this range [16]. Since distance maps have low dimensionality and are invariant to the rotation, or translation, of proteins, parameter calculation and learning becomes efficient [11], which is desirable for artificial intelligence models.

2.3   Experimental Design

The dataset has a total of 834 images, being partitioned for training, validation end test. 600 images were used for model training, 400 from the *closed* class and 200 from the *open* class. For validation, 150 images were used, 100 from the *closed* class and 50 from the *open* class. For the test set, 84 distance maps were previously separated, with the samples of each class distributed equally. To train and validate the model, the dataset was partitioned using a subset of these data for training, and validating the model in the complementary subset. This technique is known as cross-validation (CV), and it may vary according to the application [13].

However, an alternative version of this approach was used, known as *k-fold cross-validation* [29]. Basically, the technique consists of randomly partitioning the training set into k mutually exclusive subsets of the same size (n/k), where n is the total training samples. A subset is used for validation and the remaining $k-1$ are used for parameter estimation. This process is performed k times by circularly toggling the validation subset. The performance is estimated based on the average of the k error rates corresponding to each of the [13] partitions. In this case, a $k=5$ was used because, in this way, it is possible to guarantee that $\gamma \geq 0.1$, often recommended and effective in most applications [13].

Since this is a binary classification problem, the objective is to decide in which class a new observation belongs among two possible classes (*open* or *closed*). We chose to use accuracy and error to evaluate model quality, often used in binary classification problems. Precision and recall were also calculated, as we want to know the correctness rate of the model for each of the problem classes [36]. We computed the average for each metric after 5-fold CV. Adaptive Moment Estimation (Adam) was used as an optimizer [21] and, the model was trained during 100 epochs.

3   Results

The Table 1 presents the model validation results and the error rate for each partition. The model selection, by cross-validation, was based on an approach whose objective is *minimization of generalization error* [17]. Furthermore, both precision and recall suggest that the model tends to successfully discriminate both classes, even considering that the dataset is quite unbalanced [36].

For some partitions (as can be seen in the Table 1), an increase in error is directly related to a lower accuracy rate, and model precision, especially for the

Table 1. Validation performance metrics (%) obtained from 5-fold CV

Subset	Accuracy	Error	Precision	Recall
1	96.00	15.01	94.34	94.00
2	85.33	37.16	70.00	98.00
3	88.00	29.96	79.63	86.00
4	95.33	19.72	92.16	94.00
5	88.24	9.70	85.41	77.99
90.58	**22.31**	**84.32**	**89.8**

class that has few images. However, model performance improves when averaged after cross-validation, that is, considering all partitions.

In Fig. 4 presents the *learning curve* of the model after 5-fold CV. It is possible to observe an alignment between the results (training and validation), as expected from the model learning process. The validation error tends to decrease along the training epochs, indicating the non-occurrence of *overfitting* [13].

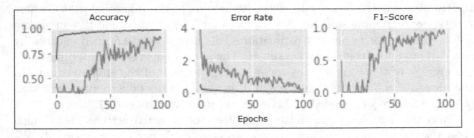

Fig. 4. ConvNet learning curves after 5-fold CV. It is possible to observe the accuracy, error rate and F1-score in the training (*blue* curve) and validation (*red* curve). Validation was performed over 100 epochs. (Color figure online)

In this sense, the use of techniques such as *dropout* [32] and *batch normalization* [18], in addition to removing noise in the input data, reduce the possibility of *overfitting*. The error rate stabilizes at 20% from the seventieth epoch, when it no longer undergoes significant changes. It is possible that the increase in the number of epochs would not bring more gains to the model.

The calculation of the F1-Score was also performed, that is, a harmonic average between precision and recall, which helps in the evaluation of the *model's discriminative capacity* [36]. The F1-Score indicates that the model can distinguish conformations, represented by patterns in distance maps Fig. 4.

Finally, the model was evaluated based on the performance achieved for the new structures presented, which constitute the test set. The *test data* correspond to variants of the original strain, with a structure very similar to those used to train the model. As a result, the model obtained an accuracy of 71.7%, with an error of 28.2%. Precision and recall reached 68.18% and 78.94%, respectively.

4 Conclusions

The development of models based on deep learning is not a trivial task requiring data pre-processing, noise removal, cross validation, etc. Since biological data are usually highly complex (incompleteness, dimensionality, noise, etc.), finding patterns in these data poses significant challenges to obtaining a predictive model capable of discriminating the classes of the problem, with high precision.

The problem addressed sought to identify large scale conformational changes, related to S proteins of SARS-CoV-2, through distance maps and ConvNets. The problem is currently relevant, as it is related to the development of bioinformatics techniques capable of raising the level of understanding about SARS-CoV-2.

Based on the results obtained for new structures (precision equal to 68.18%, and recall to 78.94%), it is concluded that the model was able to distinguish conformations referring to the open and closed states of the S protein, with a good precision. However, as these results reflect the performance of the trained model, it is still possible to adjust hyperparameters that minimize the influence of some overfitting. As future perspectives, it is intended to assess whether models based on ConvNets can identify even more subtle conformational changes, such as the impacts of a point mutation.

Acknowledgement. The authors thank the funding agencies: Coordenação de Aperfeiçoamento de Pessoal de Nível Superior (CAPES), Fundação de Amparo á Pesquisa do Estado de Minas Gerais (FAPEMIG), and Conselho Nacional de Desenvolvimento Científico e Tecnológico (CNPq).

References

1. Abadi, M., et al.: TensorFlow: large-scale machine learning on heterogeneous distributed systems, pp. 1–16 (2016). https://doi.org/10.48550/arXiv.1603.04467
2. AlQuraishi, M.: AlphaFold at CASP13. Bioinformatics **35**(22), 4862–4865 (2019). https://doi.org/10.1093/bioinformatics/btz422
3. Anishchenko, I., et al.: De novo protein design by deep network hallucination. Nature **600**, 547–552 (2020). https://doi.org/10.1038/s41586-021-04184-w
4. Baek, M., et al.: Accurate prediction of protein structures and interactions using a three-track neural network. Science **373**(6557), 871–876 (2021). https://doi.org/10.1126/science.abj8754
5. Bengio, Y., Courville, A., Vincent, P.: Representation learning: a review and new perspectives. IEEE Trans. Pattern Anal. Mach. Intell. **35**(8), 1798–1828 (2013). https://doi.org/10.1109/TPAMI.2013.50
6. Berg, J.M., Tymoczko, J.L., Stryer, L.: Biochemistry. W.H. Freeman (2002)
7. Berman, H.: The protein data bank. Nucleic Acids Res. **28**(1), 235–242 (2000). https://doi.org/10.1093/nar/28.1.235

8. Chicco, D., Heider, D., Facchiano, A.: Editorial: artificial intelligence bioinformatics: development and application of tools for omics and inter-omics studies. Front. Genet. **11** (2020). https://doi.org/10.3389/fgene.2020.00309
9. Chollet, F., et al.: Keras (2015). Keras
10. Chollet, F.: Deep Learning with Python. Manning, 4th edn. (2021)
11. Defresne, M., Barbe, S., Schiex, T.: Protein design with deep learning. Int. J. Mol. Sci. **22**(21), 11741 (2021). https://doi.org/10.3390/ijms222111741
12. Dong, E., Du, H., Gardner, L.: An interactive web-based dashboard to track COVID-19 in real time. Lancet Inf. Dis. **20**(5), 533–534 (2020). https://doi.org/10.1016/S1473-3099(20)30120-1
13. Duda, R., Hart, P., Stork, G.: Pattern Classification, 2nd edn. Wiley, New York (2001)
14. Gao, W., Mahajan, S., Sulam, J., Gray, J.: Deep learning in protein structural modeling and design. Patterns **1** (2020). https://doi.org/10.1016/j.patter.2020.100142
15. Goodsell, D., Dutta, S., Zardecki, C., Voigt, M., Berman, H., Burley, S.: The RCSB PDB molecule of the month: inspiring a molecular view of biology. PLoS Biol. **13**(5), 1–12 (2015). https://doi.org/10.1371/journal.pbio.1002140
16. Goodfellow, I., Bengio, Y., Courville, A.: Deep Learning. Adaptive Computation and Machine Learning, MIT Press, Cambridge (2016)
17. Haykin, S.: Neural Networks - A Comprehensive Foundation. Pearson Prentice Hall, Upper Saddle River (2001)
18. Ioffe, S., Szegedy, C.: Batch normalization: accelerating deep network training by reducing internal covariate shift. In: Proceedings of the 32nd International Conference on International Conference on Machine Learning, ICML 2015, pp. 448–456 (2015). https://doi.org/10.48550/arXiv.1502.03167
19. Iyer, M., Jaroszewski, L., Sedova, M., Godzik, A.: What the protein data bank tells us about the evolutionary conservation of protein conformational diversity. Protein Sci. **31**(7) (2022). https://doi.org/10.1002/pro.4325
20. Jumper, J., et al.: Highly accurate protein structure prediction with AlphaFold. Nature **596**, 583–589 (2021). https://doi.org/10.1038/s41586-021-03819-2
21. Kingma, D., Ba, J.: Adam: a method for stochastic optimization. Published as a Conference Paper at the 3rd International Conference for Learning Representations, San Diego (2015). https://doi.org/10.48550/arXiv.1412.6980
22. Kloczkowski, A., et al.: Distance matrix-based approach to protein structure prediction. J. Struct. Funct. Genomics **10**(1), 67–81 (2009). https://doi.org/10.1007/s10969-009-9062-2
23. Leach, A.: Molecular Modelling: Principles and Applications. Prentice Hall, New York (2001)
24. LeCun, Y., et al.: Backpropagation applied to handwritten zip code recognition. Neural Comput. **1**(4), 541–551 (1989)
25. Li, Q., et al.: Early transmission dynamics in Wuhan, China, of novel coronavirus-infected pneumonia. N. Engl. J. Med. **382**(13), 1199–1207 (2020). https://doi.org/10.1056/NEJMoa2001316
26. Min, S., Lee, B., Yoon, S.: Deep learning in bioinformatics. Brief. Bioinform. **18**(5), 851–869 (2017). https://doi.org/10.1093/bib/bbw068
27. Mishkin, D., Sergievskiy, N., Matas, J.: Systematic evaluation of CNN advances on the ImageNet. https://doi.org/10.1016/j.cviu.2017.05.007
28. Mitchell, T.: Machine Learning. McGraw-Hill, New York (1997)
29. Mosteller, F., Tukey, J.: Data analysis, including statistics. In: Lindzey, G., Aronson, E. (eds.) Revised Handbook of Social Psychology, vol. 2, pp. 80–203 (1968)

30. Narayanan, A., Keedwell, E., Olsson, B.: Artificial intelligence techniques for bioinformatics. Appl. Bioinform. **1**, 191–222 (2002)
31. Nicolas, J.: Artificial intelligence and bioinformatics. In: Marquis, P., Papini, O., Prade, H. (eds.) A Guided Tour of Artificial Intelligence Research, pp. 209–264. Springer, Cham (2020). https://doi.org/10.1007/978-3-030-06170-8_7
32. Srivastava, N., Hinton, G., Krizhevsky, A., Sutskever, I., Salakhutdinov, R.: Dropout: a simple way to prevent neural networks from overfitting. J. Mach. Learn. Res. **15**(56), 1929–1958 (2014). https://doi.org/10.5555/2627435.2670313
33. Torrisi, M., Pollastri, G., Le, Q.: Deep learning methods in protein structure prediction. Comput. Struct. Biotechnol. J. **18**, 1301–1310 (2020)
34. Yang, J., Anishchenko, I., Park, H., Peng, Z., Ovchinnikov, S., Baker, D.: Improved protein structure prediction using predicted interresidue orientations. Proc. Natl. Acad. Sci. **117**(3), 1496–1503 (2020). https://doi.org/10.1073/pnas.1914677117
35. Walls, A., Park, Y., Tortorici, M., Wall, A., McGuire, A., Veesler, D.: Structure, function, and antigenicity of the SARS-CoV-2 spike glycoprotein. Cell **181**(2), 281–292 (2020). https://doi.org/10.1016/j.cell.2020.02.058
36. Webb, A., Copsey, K.: Statistical Pattern Recognition. Wiley, New York (2011)
37. Wu, F., et al.: A new coronavirus associated with human respiratory disease in China. Nature **579**, 265–269 (2020). https://doi.org/10.1038/s41586-020-2008-3
38. Zhang, Y.: TM-align: a protein structure alignment algorithm based on the TM-score. Nucleic Acids Res. **33**(7), 2302–2309 (2005)
39. Zhu, N., et al.: A novel coronavirus from patients with pneumonia in China, 2019. N. Engl. J. Med. **382**(8), 727–733 (2020). https://doi.org/10.1056/NEJMoa2001017

Clustering Analysis Indicates Genes Involved in Progesterone-Induced Oxidative Stress in Pancreatic Beta Cells: Insights to Understanding Gestational Diabetes

Lara Marinelli Dativo dos Santos[✉], Patricia Rufino Oliveira, and Anna Karenina Azevedo Martins[iD]

School of Arts, Sciences and Humanities, Avenida Arlindo Bettio, 1000, São Paulo, SP 03828-000, Brazil
{lara.santos,proliveira,karenina}@usp.br

Abstract. Clustering analysis in gene expression data has been shown to be useful for understanding gene function, gene regulation, and cell processes and subtypes. Due to the wide availability of techniques for this task, the choice of an appropriate method is critical. Trying to mitigate this problem, Saelens and coauthors performed, in 2018, a benchmark study based on external validation indices. The present work proposes an extension of this analysis by including internal indices and applying it in a study case to investigate gestational diabetes through experiments on microarray data of pancreatic beta cells submitted to supra-pharmacological doses of progesterone. The results of the clustering method selected by the proposed extension have shown to be helpful in an enrichment analysis that identified TXNIP gene as relevant for future work aiming at understanding in more details the gestational diabetes phenomena.

Keywords: Gestational diabetes · Clustering methods · Gene expression data analysis

1 Introduction

Clustering analysis on gene expression data is a widely used technique to help understand gene function, gene regulation, and cell processes and subtypes and has been consistently applied to identify and analyze various pathologies such as cancer, malaria, and tuberculosis [5]. Such analysis tries to identify the function of genes based on a principle known in the literature as *guilt by association* [8,20], which establishes that genes with similar functions tend to be clustered together. The fact that certain genes have been allocated to the same group,

The authors thank to the high performance computing resources of University of São Paulo (https://hpc.usp.br/).

which in the context of gene expression is called a module, may indicate that such genes are related to the same cellular processes, that they present co-regulation or that they share a common mechanism. Furthermore, the detected groups tend to be significantly enriched with specific functional categories, a fact that can be exploited to infer the function of genes in a given context [5,11]. The clustering of gene expression data can be validated in two main ways: through internal indexes, inherent to the cluster itself, or external indexes, based on the agreement between the obtained clusters and a reference cluster [11,21].

Due to the existence of a large number of clustering techniques, choosing the most suitable method for a given application becomes a challenge. In an attempt to mitigate this problem, the work in [27] presents an overview of the character-istics and performance of clustering techniques applied to gene expression data and proposes a *benchmark* strategy for carrying out comparative studies. In such work, 49 methods were applied to nine gene expression datasets and evaluated according to external indexes, so that a score was assigned to each method based on the agreement between modules identified in the experiments and reference modules established by regulatory networks. According to the scoring scheme developed by the authors, the method that obtained the best performance was that based on independent component analysis (ICA).

Several later studies [2,15,24,28,30] selected the ICA technique for their experiments based on the results presented in [27]. The problem here is that this choice is only supported by external evaluation results, which do not take into account the nature of the problem and the inherent clusters. In fact, as stated in [31], in a real-world scenario (generally without a reference cluster) the researcher who performs a clustering task on a new dataset has only the availability of internal validation indices. The present work intends to extend the discussion on the *benchmark* developed in [27] to include internal indices and the interpretation of the clusters detected through functional analysis studies. As a case study, a microarray dataset of pancreatic beta cells submitted to progesterone [19] will be analyzed.

The remainder of this paper is organized as follows. Section 2 introduces the problem definition. Methods and background are described in Sect. 3. Experi-mental results are discussed in Sect. 4. Finally, Sect. 5 presents some conclusions and future work.

2 Problem Definition

Due to the increasing pharmacological use of progestogens throughout preg-nancy for the prevention of preterm birth [22], understanding the relationship between these hormones and gestational diabetes requires attention. From *in vitro* experiments with the Rinm5f cell line, it was found that progesterone was able to induce the oxidation and death of pancreatic beta cells, which are insulin producers [19].

Pancreatic beta cell death is associated with both type I and type II diabetes [25], but still needs to be better understood in the context of gestational dia-betes. To investigate this problem, microarray experiments were conducted by submitting Rinm5f cell line to progesterone at three doses (0.1 μM, 1 μM and 100 μM) and two time points (6 h and 24 h), which were analyzed in this work.

3 Methods

The experiments can be subdivided into four steps: (i) reproduction of the experiments in [27]; (ii) extension of the discussion by calculating and proposing a score based on validation indices; (iii) reproduction of all experiments including the dataset of pancreatic beta cells subjected to progesterone; (iv) analysis of the best results for the dataset of pancreatic beta cells according the internal and external validation indices.

3.1 Clustering Methods

Five clustering methods were analyzed, including the best evaluated in [27]: k-means [17], agglomerative hierarchical clustering [10], spectral bi-clustering [12], ICA z-score [9], meanshift [3], and random clustering (baseline) [27].

3.2 Reference Modules

External validation measures are calculated by comparing the obtained clusters with reference modules. For the benchmark experiments, such modules were defined according to three different criteria: (i) *minimal* (genes that have at least one element in common), (ii) *strict* (genes that have exactly the same regulators in common), and (iii) *interconnected subgraphs* (based on the construction of graphs from regulatory networks). For the dataset studied in this work, given the absence of regulatory networks, the reference modules were obtained through the application of the WGCNA algorithm [14] and also through the construction of modules from [29], applying *minimum* and *strict* co-regulation criteria.

3.3 Validation Indices

Validation measures are classified into external and internal, as follows:

- **External indices:** Four different external validation measures were used (Recovery, Relevance, Recall and Precision), and then normalized and combined through the harmonic mean in a measure called F1rprr, according to the scoring scheme proposed in [27].
- **Internal indices:** Internal indices are based on two main concepts: cohesion, which refers to how far the items in the same group are (intra-group distance); and separation, which aims to quantify the distance between groups. Intra-group and inter-group distances can be calculated using a given distance criterion. According to the internal indices, the best evaluated clustering for a given dataset is the one that minimizes the intra-group distance and maximizes the distance between groups. The indices that were used are listed in the Table 1.

Table 1. Internal validity indexes

Index	Best if...	Range
Dunn [7]	High	$[-\infty, +\infty]$
Davies-Bouldin [6]	Low	$[-\infty, +\infty]$
Silhouette [26]	High	$[-1, +1]$

3.4   Functional Enrichment Analysis

The tools Enrichr [13] and that developed in [18] where used for the functional enrichment analysis. The databases DisGeNET [23], GO Biological Process 2021, GO Cellular Component 2021, GO Molecular Function 2021 [4] were also selected for this task.

3.5   Microarray Experiments

The dataset containing information from 89 genes was obtained in experiments conducted by one of the authors of this paper, and presented by the first time here, using microarrays with *RT2 Profiler PCR Arrays kit*, from the manufacturer *Qiagen*[1]. Cells were subjected to concentrations of 0.1 μM, 1 μM and 100 μM and expressions relative to the control group were collected at 6 h and 24 h from the beginning of the experiment, resulting in six measurements. For visualization purposes, the experiments (0.1 μM, 6 h), (0.1 μM, 24 h),(1 μM, 6 h),(1 μM, 24 h),(100 μM, 6 h), (100 μM, 6 h) are referenced by I, II, III, IV, V and VI, respectively. Hereafter, such dataset is named by 'progesterone dataset'.

4   Experimental Results

4.1   Module Detection Experiments

Initially, experiments were conducted for the benchmark as proposed in [27], but now including the progesterone dataset[2]. To avoid parameter overfitting on particular datasets, we first optimized the parameters for all methods and datasets using a grid search and calculated the training scores. Next, such parameters were applied to assess the performance of each method on all different datasets (excluding the one used for training), obtaining the test scores, as proposed in [27]. These results are presented in Fig. 1.

According to external indices scores, agglomerative clustering, meanshift and k-means were the best evaluated methods for progesterone dataset, when considering tests scores. Furthermore, it can be noted that ICA z-score was well evaluated according to training scores, but the same does not occur for testing

---

[1] https://www.qiagen.com/us/.

[2] The human datasets were excluded since the authors used a different criteria for module definition, called 'regulatory circuits'.

scores. Ideed, after functional enrichment analysis, we also verified that ICA did not show a good performance for these data. This fact highlights the importance of the training-and-test strategy.

After these preliminary results, we extended the discussion by also calculating internal indeces and expanding the score strategy, adding three different scores: (i) Dunn score; (ii) Davies-Bouldin score; and (iii) silhouette score.

For $E.\ coli$ data, Dunn scores (Fig. 2) are in agreement with external validation scores, presenting the same order for the top scores. Considering the progesterone dataset, we find that the best evaluated method was k-means, which was the third method according to external validating indices.

From Davies-Boundin score perspective (Fig. 3), meanshift and k-means were the best evaluated methods. However, meanshift results does not contribute to the problems comprehension, since all genes were allocated in one single cluster. Finally, silhouette scores (Fig. 4) also indicated k-means as the best method.

After this investigation, we selected k-means with its best clustering configuration ($k = 4$) for carring out functional enrichment analysis. The modules that resulted from these clustering are shown in the Fig. 5. We can see that there are two generally overexpressed modules (Figs. 5a and 5c) and two underexpressed (Figs. 5b 5d).

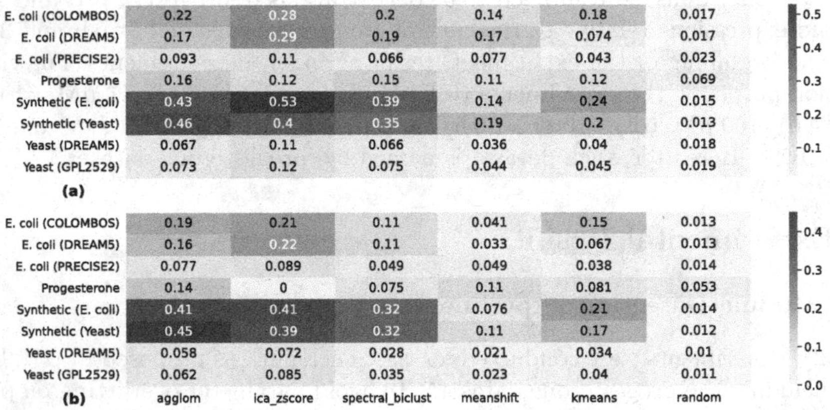

	agglom	ica_zscore	spectral_biclust	meanshift	kmeans	random
E. coli (COLOMBOS)	0.22	0.28	0.2	0.14	0.18	0.017
E. coli (DREAM5)	0.17	0.29	0.19	0.17	0.074	0.017
E. coli (PRECISE2)	0.093	0.11	0.066	0.077	0.043	0.023
Progesterone	0.16	0.12	0.15	0.11	0.12	0.069
Synthetic (E. coli)	0.43	0.53	0.39	0.14	0.24	0.015
Synthetic (Yeast)	0.46	0.4	0.35	0.19	0.2	0.013
Yeast (DREAM5)	0.067	0.11	0.066	0.036	0.04	0.018
Yeast (GPL2529)	0.073	0.12	0.075	0.044	0.045	0.019
(a)						
E. coli (COLOMBOS)	0.19	0.21	0.11	0.041	0.15	0.013
E. coli (DREAM5)	0.16	0.22	0.11	0.033	0.067	0.013
E. coli (PRECISE2)	0.077	0.089	0.049	0.049	0.038	0.014
Progesterone	0.14	0	0.075	0.11	0.081	0.053
Synthetic (E. coli)	0.41	0.41	0.32	0.076	0.21	0.014
Synthetic (Yeast)	0.45	0.39	0.32	0.11	0.17	0.012
Yeast (DREAM5)	0.058	0.072	0.028	0.021	0.034	0.01
Yeast (GPL2529)	0.062	0.085	0.035	0.023	0.04	0.011
(b)	agglom	ica_zscore	spectral_biclust	meanshift	kmeans	random

Fig. 1. Training (a) and (b) test scores for each dataset and method

4.2   Functional Enrichment Analysis

Functional enrichment analysis was performed using various datasets available on Enrichr and the most relevant results to the problem are discussed here. For understanding purposes, in next sections the modules are referred as: (i) module A (Fig. 5a); (ii) module B (Fig. 5b); (iii) module C (Fig. 5c); and (iv) module D (Fig. 5d).

(a)

	agglom	ica_zscore	spectral_biclust	meanshift	kmeans	random
E. coli (COLOMBOS)	0.19	0.17	0.23	0.44	0.26	0.16
E. coli (DREAM5)	0.037	0.026	0.041	0.16	0.13	0.021
E. coli (PRECISE2)	0.058	0.03	0.054	0.095	0.075	0.013
Progesterone	0.0094	0.0047	0.0037	0.0088	1.5	0.003
Synthetic (E. coli)	0.25	0.22	0.28	0.7	0.38	0.21
Synthetic (Yeast)	0.19	0.18	0.22	0.39	0.37	0.17
Yeast (GPL2529)	0.052	0.038	0.052	0.32	0.11	0.025
Yeast (DREAM5)	0.051	0.023	0.049	0.16	0.096	0.021

(b)

	agglom	ica_zscore	spectral_biclust	meanshift	kmeans	random
E. coli (COLOMBOS)	0.18	0.14	0.2	0.087	0.23	0.15
E. coli (DREAM5)	0.034	0.012	0.033	0.043	0.11	0.02
E. coli (PRECISE2)	0.052	0.01	0.041	0.047	0.054	0.011
Progesterone	0.0051	0	0.0028	0	0.065	0.0027
Synthetic (E. coli)	0.24	0.17	0.25	0.067	0.36	0.2
Synthetic (Yeast)	0.17	0.14	0.19	0	0.32	0.16
Yeast (GPL2529)	0.048	0.0092	0.044	0.043	0.09	0.024
Yeast (DREAM5)	0.043	0.012	0.032	0.037	0.083	0.021

Fig. 2. Training (a) and (b) test Dunn scores for each dataset and method

(a)

	agglom	ica_zscore	spectral_biclust	meanshift	kmeans	random
E. coli (COLOMBOS)	4.2	11	4.8	1.5	3.2	29
E. coli (DREAM5)	4.6	17	5.5	5.4	2	45
E. coli (PRECISE2)	3.5	10	4.9	2.5	2	46
Progesterone	2.1	8.8	4.1	1.8	1	7.9
Synthetic (E. coli)	5.7	7.2	5.6	0.86	3.6	25
Synthetic (Yeast)	5.6	7.9	5.2	1.1	3.4	27
Yeast (GPL2529)	4.4	16	5.2	3.5	2.4	61
Yeast (DREAM5)	4.7	19	7.1	4.6	2.2	53

(b)

	agglom	ica_zscore	spectral_biclust	meanshift	kmeans	random
E. coli (COLOMBOS)	3.6	10	4.1	0.46	2.7	20
E. coli (DREAM5)	4.2	10	5.1	0.77	1.9	30
E. coli (PRECISE2)	3.3	9.7	4.4	0.5	1.9	32
Progesterone	1.7	4.2	3.5	0	0.53	7.3
Synthetic (E. coli)	4.3	6.7	4.6	0.12	3.2	18
Synthetic (Yeast)	4.3	7	4.5	0.17	3.2	19
Yeast (GPL2529)	4.1	15	4.7	1.2	2.2	40
Yeast (DREAM5)	4.5	16	6.3	0	2.1	33

Fig. 3. Training (a) and (b) test Davies-Bouldin scores for each dataset and method

(a)

	agglom	ica_zscore	spectral_biclust	meanshift	kmeans	random
E. coli (COLOMBOS)	-0.012	-0.026	0.019	0.01	0.05	-0.012
E. coli (DREAM5)	-0.12	-0.15	-0.021	4.2e-05	0.1	-0.018
E. coli (PRECISE2)	-0.092	-0.051	-0.017	0.033	0.045	-0.015
Progesterone	0.17	-0.15	0.092	0.17	0.87	-0.097
Synthetic (E. coli)	0.0072	-0.057	0.0034	0.066	0.025	-0.011
Synthetic (Yeast)	0.029	-0.075	0.006	0.0025	0.031	-0.01
Yeast (GPL2529)	-0.094	-0.17	-0.026	0.047	0.094	-0.025
Yeast (DREAM5)	-0.14	-0.15	-0.038	-0.0037	0.095	-0.02

(b)

	agglom	ica_zscore	spectral_biclust	meanshift	kmeans	random
E. coli (COLOMBOS)	-0.06	-0.056	0.013	-0.001	0.04	-0.036
E. coli (DREAM5)	-0.22	-0.21	-0.054	-0.0035	0.077	-0.08
E. coli (PRECISE2)	-0.13	-0.17	-0.054	0.0034	0.044	-0.05
Progesterone	-0.017	-0.094	-0.34	0	0.48	-0.16
Synthetic (E. coli)	-0.016	-0.17	-0.034	0.0016	0.023	-0.047
Synthetic (Yeast)	0.0022	-0.11	-0.02	-0.00047	0.016	-0.034
Yeast (GPL2529)	-0.19	-0.23	-0.058	-0.068	0.059	-0.084
Yeast (DREAM5)	-0.22	-0.15	-0.085	-0.018	0.066	-0.071

Fig. 4. Training (a) and (b) test Silhouette scores for each dataset and method

Module A. This module was enriched through DisGeNET with Diabetes type I and type II related terms (Fig. 6). This fact is in agreement with the relation of oxidative stress in pancreatic beta cells and Diabetes. On other hand, functional enrichment with the Gene Ontology returns terms such as *cellular response to oxidative stress* and response with *cytokine-mediated signaling pathway*. Through the *guilt by association* principle, it can be inferred that the genes of this module have a role in the response to progesterone stimulus, or in anti-oxidant defense.

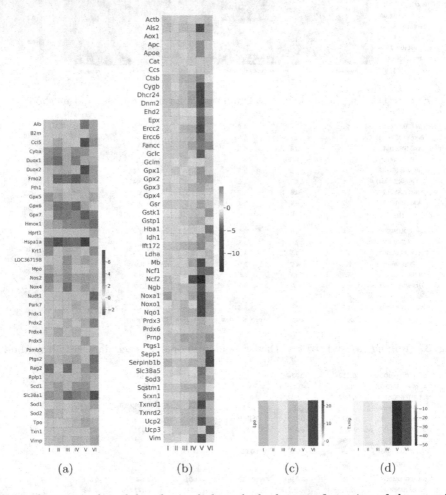

Fig. 5. Overview of modules obtained through the best configuration of the experiments. Figures (a), (b), (c) and (d) illustrate the allocation of genes for detected modules. For each module, the experiments are denoted by I, II, III, IV, V and VI. The colors represent the expression of genes during the experimental condition in relation to the control group.

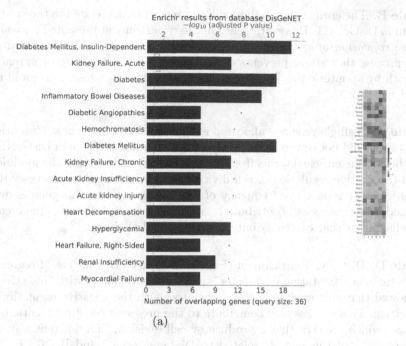

Fig. 6. Functional enrichment through the DisGeNET for module A

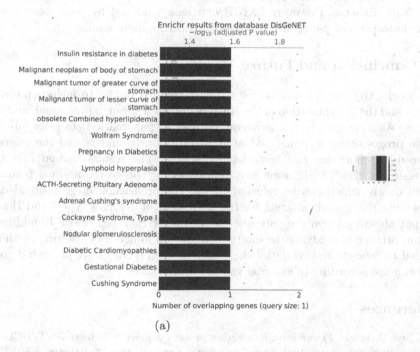

Fig. 7. Functional enrichment through the DisGeNET for module D

Module B. The enrichment analysis did not return terms related to the studied problem in DisGeNET. However, Gene Ontology enrichment presented terms like *negative regulation of apoptotic process* (or anti-apoptosis), which by definition is any process that stops, prevents or reduces the frequency, rate or extent of cell death by apoptotic process. Since this module was suppressed, it could not prevent cell death.

Module C. A single gene was allocated in this module (LPO gene). The enrichment analysis did not return terms related to the studied problem in DisGeNET. Nevertheless, the enriched terms from Gene Ontology showed that the module is related to negative regulation of cell division, which is related to processes that stop, prevent or reduce the frequency of cell division. Since this gene is over-expressed, this process can contribute to an imbalance in the mass of pancreatic beta cells, factor that can contribute to Gestational Diabetes.

Module D. DisGeNET enrichment (Fig. 7), returned terms such as "Pregnancy in diabetes" and "Gestational Diabetes", while the enrichment with Gene Ontology showed that this module is also associated with the negative regulation of cell division. This process may contribute to the preservation of pancreatic beta cell mass, which are cells that reproduce by cell division. In a different way, the over-expression of this gene is related to Diabetes type I and II [16,32]. The study of [1] points out that TXNIP deficiency was able to completely rescue mice from diabetes. However, TXNIP deficiency induced by progesterone was not able to prevent pancreatic cell death, requiring more studies.

5   Conclusion and Future Work

This work extended the *benchmark* study developed in [27] to include internal indices and the interpretation of the clusters detected through functional analysis studies. As a case study, a microarray dataset of pancreatic beta cells submitted to progesterone was analyzed in an attempt to understand the relation between this hormone and gestational diabetes. The results showed that the expression of the TXNIP gene was suppressed by progesterone in a scenario of cell death, which can be relevant to the comprehension of gestational diabetes since this gene is a target for the treatment of diabetes type I and II and was not able to prevent progesterone-induced pancreatic cell death. In addition, the importance of studying the internal validation indices of clustering methods applied to gene expression data is highlighted, even if they have presented good performance according to external validation measures.

References

1. Chen, J., et al.: Thioredoxin-interacting protein deficiency induces AKT/BCL-XL signaling and pancreatic beta-cell mass and protects against diabetes. FASEB J. **22**(10), 3581–3594 (2008)

2. Chen, R.Y., et al.: Duodenal microbiota in stunted undernourished children with enteropathy. N. Engl. J. Med. **383**(4), 321–333 (2020)
3. Comaniciu, D., Meer, P.: Mean shift: a robust approach toward feature space analysis. IEEE Trans. Pattern Anal. Mach. Intell. **24**(5), 603–619 (2002)
4. Gene Ontology Consortium: Gene ontology consortium: going forward. Nucleic Acids Res. **43**(D1), D1049–D1056 (2015)
5. Dalton, L., Ballarin, V., Brun, M.: Clustering algorithms: on learning, validation, performance, and applications to genomics. Curr. Genomics **10**(6), 430–445 (2009)
6. Davies, D.L., Bouldin, D.W.: A cluster separation measure. IEEE Trans. Pattern Anal. Mach. Intell. **2**, 224–227 (1979)
7. Dunn, J.C.: Well-separated clusters and optimal fuzzy partitions. J. Cybern. **4**(1), 95–104 (1974)
8. Eisen, M.B., Spellman, P.T., Brown, P.O., Botstein, D.: Cluster analysis and display of genome-wide expression patterns. Proc. Natl. Acad. Sci. **95**(25), 14863–14868 (1998)
9. Hyvärinen, A., Oja, E.: Independent component analysis: algorithms and applications. Neural Netw. **13**(4–5), 411–430 (2000)
10. Jain, A.K., Dubes, R.C.: Algorithms for Clustering Data. Prentice-Hall Inc., Hoboken (1988)
11. Jiang, D., Tang, C., Zhang, A.: Cluster analysis for gene expression data: a survey. IEEE Trans. Knowl. Data Eng. **16**(11), 1370–1386 (2004). https://doi.org/10.1109/TKDE.2004.68
12. Kluger, Y., Basri, R., Chang, J.T., Gerstein, M.: Spectral biclustering of microarray data: coclustering genes and conditions. Genome Res. **13**(4), 703–716 (2003)
13. Kuleshov, M.V., et al.: Enrichr: a comprehensive gene set enrichment analysis web server 2016 update. Nucleic Acids Res. **44**(W1), W90–W97 (2016)
14. Langfelder, P., Horvath, S.: WGCNA: an R package for weighted correlation network analysis. BMC Bioinform. **9**(1), 1–13 (2008)
15. Lawlor, M.A., Cao, W., Ellison, C.E.: A transposon expression burst accompanies the activation of Y-chromosome fertility genes during drosophila spermatogenesis. Nat. Commun. **12**(1), 1–12 (2021)
16. Lei, Z., et al.: TXNIP deficiency promotes β-cell proliferation in the HFD-induced obesity mouse model. Endocrine Connections **11**(4) (2022)
17. Lloyd, S.: Least squares quantization in PCM. IEEE Trans. Inf. Theory **28**(2), 129–137 (1982)
18. Luebbert, L., Pachter, L.: Efficient querying of genomic reference databases with gget. bioRxiv (2022)
19. Nunes, V.A., et al.: Progesterone induces apoptosis of insulin-secreting cells: insights into the molecular mechanism. J. Endocrinol. **221**(2), 273–284 (2014)
20. Oliver, S.: Guilt-by-association goes global. Nature **403**(6770), 601–602 (2000)
21. Oyelade, J., et al.: Clustering algorithms: their application to gene expression data. Bioinform. Biol. Insights **10**, BBI-S38316 (2016)
22. Pergialiotis, V., Bellos, I., Hatziagelaki, E., Antsaklis, A., Loutradis, D., Daskalakis, G.: Progestogens for the prevention of preterm birth and risk of developing gestational diabetes mellitus: a meta-analysis. Am. J. Obstet. Gynecol. **221**(5), 429–436 (2019)
23. Piñero, J., et al.: The DisGeNET knowledge platform for disease genomics: 2019 update. Nucleic Acids Res. **48**(D1), D845–D855 (2020)
24. Poudel, S., et al.: Revealing 29 sets of independently modulated genes in staphylococcus aureus, their regulators, and role in key physiological response. Proc. Natl. Acad. Sci. **117**(29), 17228–17239 (2020)

25. Rojas, J., et al.: Pancreatic beta cell death: novel potential mechanisms in diabetes therapy. J. Diab. Res. **2018** (2018)
26. Rousseeuw, P.J.: Silhouettes: a graphical aid to the interpretation and validation of cluster analysis. J. Comput. Appl. Math. **20**, 53–65 (1987)
27. Saelens, W., Cannoodt, R., Saeys, Y.: A comprehensive evaluation of module detection methods for gene expression data. Nat. Commun. **9**(1), 1–12 (2018)
28. Sastry, A.V., Hu, A., Heckmann, D., Poudel, S., Kavvas, E., Palsson, B.O.: Independent component analysis recovers consistent regulatory signals from disparate datasets. PLoS Comput. Biol. **17**(2), e1008647 (2021)
29. Shannon, P., et al.: Cytoscape: a software environment for integrated models of biomolecular interaction networks. Genome Res. **13**(11), 2498–2504 (2003)
30. Tan, J., et al.: Independent component analysis of E. coli's transcriptome reveals the cellular processes that respond to heterologous gene expression. Metabolic Eng. **61**, 360–368 (2020)
31. Wiwie, C., Baumbach, J., Röttger, R.: Comparing the performance of biomedical clustering methods. Nat. Methods **12**(11), 1033–1038 (2015)
32. Wondafrash, D.Z., Nire'a, A.T., Tafere, G.G., Desta, D.M., Berhe, D.A., Zewdie, K.A.: Thioredoxin-interacting protein as a novel potential therapeutic target in diabetes mellitus and its underlying complications. Diab. Metab. Syndr. Obes.: Targets Ther. **13**, 43 (2020)

An External Memory Approach for Large Genome *De Novo* Assembly

Elvismary Molina de Armas(✉)⬤ and Sérgio Lifschitz⬤

Depto. Informática, PUC-Rio, Rio de Janeiro, Brazil
{earmas,sergio}@inf.puc-rio.br

Abstract. De novo genome assembly of sequenced reads is a fundamental problem in bioinformatics. When there is no reference genome sequence to guide the process, many assemblers programs consider using the *de Bruijn* Graph data structure to improve performances. However, the construction of such a graph has a high computational cost, mainly due to internal RAM consumption in the presence of very large and repeated read datasets. Building a *de Bruijn* Graph relies on a broad set of k-mers. Some existing approaches use external memory processing to make it feasible. This work proposes an approach for constructing the *de Bruijn* graph that does not generate all k-mers during the execution. An external memory processing allows reducing the high number of duplicate k-mers and, consequently, reduces the total number of k-mers that incur on the number of I/O operations. Some practical experiments are presented, showing the solution's viability and its improvements over other common assemblers in the literature. Our solution reduces the computational requirements and enables execution feasibility.

Keywords: *de Bruijn* graph · k-mer · External memory processing · *De novo* assembly

1 Introduction

Next-generation sequencing (NGS) technologies have brought rapid progress for the biological research area. Nevertheless, the genome assembly problem continues to be a challenge since we need to reconstruct a whole genome by joining a vast amount of short reads.

Assembly algorithms and their implementations are typically complex. They could require high-performance computing platforms for large genomes. Algorithmic success can depend on pragmatic engineering and heuristics formulated by empirically derived rules of thumb.

Since fragments of DNA are broken in random positions, and sequencer machines do not have a 100% of accuracy, it is needed to increase the sequencing coverage. The coverage is measured as a function of the average number of reads covering a position in the genome.

Given the pieces taking from unknown positions and the great coverage, a high redundancy level is generated in the fragments. The number of reads could

N. M. Scherer and R. C. de Melo-Minardi (Eds.): BSB 2022, LNBI 13523, pp. 79–90, 2022.
https://doi.org/10.1007/978-3-031-21175-1_9

be hundreds of millions; thus, the total volume of data may reach tens or even hundreds of GB. For *de novo* assembly [15], without a known reference genome, the complexity is higher.

Some successful approaches are based on the use of *de Bruijn* graphs (see [3] and [14]). However, the construction and use of *de Bruijn* graphs (DBG) demand a large amount of main memory and execution time because of the large number of elements (nodes and edges) to process [3,6,12,13].

In a DBG, unique k-mers constitute nodes, and an edge is set between two nodes when the k-mers of those nodes overlaps k-1 symbols in at least one read. The total number of k-mers present in one read (not only distinct k-mers), with length m, is equal to $m - k + 1$, while the total number of k-mers present in n reads is $(m - k + 1) * n$. As n in practice is very large, the number of k-mers is even larger and computational limits may be reached.

This papers details a new approach for *de Bruijn* graph construction in *de novo* genome assembly that avoids generating all k-mers. We give a detailed theoretical explanation of how our approach becomes feasible for external memory optimization. Furthermore, we show some practical experiments on real data that illustrate and validate our ideas.

This text is structured as follows: in the next section, Sect. 2, we describe the related works within the context of this research. The basics of the new approach are shown in Sect. 3. Next, in Sect. 4, a detailed analysis of the external memory processing is exposed. Finally, tests and results are shown in Sect. 5 together with a comparison with other commons assemblers.

2   Related Works

Few works have been found that encourage the *de Bruijn* graph construction using an external memory processing [4,5,11,13], and [2,8,9,16]. They might be classified into four categories: out-of-core sorting, second based on k-mers partitioning and disk distribution, third based on memory frugal and partition in disk, and the last one based on the construction of the graph embedded into a relational database management system (RDBMS).

K-mers partitioning and disk distribution are common approach for the Minimum Substring Partition (MSP) approach [13], BCALM1 [4] and BCALM2 [5]. The distributed processing firstly distributes all k-mers into disk partitions (not disjoint partition for all cases), then processes each partition individually in the main memory, for later merges the results to build a DBG. In the case of (MSP) approach, the partitions are made based on the minimum p-substring of the k-mers, allowing consecutive k-mers to be distributed in the same partition, decreasing the number of I/O operations. BCALM1 and BCALM2 solutions, are different from MSP in the sense that they have the goal to obtain compacted graphs by the compression of all its maximal non-branching paths. They partition using a hash function in DSK (for BCALM1) and the concept of minimizers. Their pipes allow that same k-mers to be distributed in more than one file partition, and hash functions used do not warrant balanced partition file sizes, affecting the I/O throughput.

Finally, a group of published works [2,8,9,16], test the viability of using Relational Database Management System (RDBMS) to management the main and external memory interchange in the construction of the graph. In [8] was described the k-mer mapping process as part of the DBG construction, while a case study was implemented based on the Velvet assembler algorithm using PostgreSQL. Later, an ad-hoc cost model to measure the performance gained using different index structures is presented in [8], while in [9] is exposed an study of indexes like B+-tree, hash over k-mer in [9] and over k-mer p-minimum substring in [2]. As a distinguishing feature, the use of DBMS to manage the I/O operations in the mapping k-mer process allows incremental processing without reprocessing and recovery from failures [16], and the trust of a robust and very well tested system. However, some optimizations are needed to improve the run-time given by index evaluation, which could still be considered high.

In summarizing, the external memory DBG construction approaches studied, initially considering the total number of k-mers, for later obtaining the vertices of the graph (unique k-mers) and corresponding edges. Using the total number of k-mers in external memory operations implies maintaining a high level of redundancy. Consequently, a high available memory resource (main or external) and a more significant number of I/O operations are needed.

3   *de Bruijn* Graph Approach Construction

In [7] was presented a novel approach to construct DBG for assembly domains, which constitutes the preliminaries of the present work. Based on the fact that k-mers are units that encapsulate a high level of redundancy, it is clear that working with the total number of k-mers implies maintaining a high level of redundancy. Consequently, a significant amount of memory (RAM or external) and I/O operations.

The approach for DBG construction for genome fragment assembly was founded on the following principles [7]:

1. *Find overlaps regions greater than k earlier* can save the corresponding memory to store the redundant information for each k-mer and redundant information for consecutive k-mer chains that are duplicated.
2. *Avoid generate all k-mers using iterative reduction steps.*
3. *Use external processing only in the last steps of the current DBG construction approaches* with a minor number of elements.

3.1   Algorithm to DBG Construction

Definition 1. *The de Bruijn graph, $G_k(V,E)$ represents overlaps between k-mers, in which:*

- *The set of vertices is defined by $V = S = \{s_1, s_2, ..., s_p\}$, where S is a set of unique k-mers over a given set of reads.*

- *The set of edges is defined by $E = \{e_1, e_2, ..., e_q\}$, where $e = (s_i, s_j)$ if and only if the $k - 1$ suffix of s_i matches exactly the $k - 1$ prefix of s_j. s_i and s_j must be adjacent k-mers in at least one read.*

Definition 2. *A dk-mer is a substring of a genome piece with specified d length (also called dimension in this work), with $d \geq k$, over the alphabet of bases $\Sigma = \{A, T, C, G\}$.*

Two dk-mers, $dk - mer_1$ and $dk - mer_2$ are adjacent if they share $k - 1$ bases ($k - 1$ suffix of the first is equal to the $k - 1$ prefix of the second) and they are adjacent in at least one read.

The Fig. 1a shows a dk-mers representation over read r and how adjacent dk-mers share $k - 1$ bases.

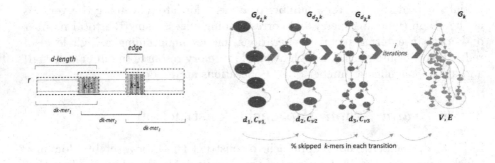

(a) dk-mers representation over read r.   (b) de Bruijn Graph generation through a iterative steps of generation of extra-compacted de Bruijn Graph and skipped k-mers reduction.

Fig. 1. *de Bruijn Graph* approach construction

Definition 3. *An extra-compacted de Bruijn Graph $G_{d,k}(V_{d,k}, E_{d,k})$ is a graph $G(V, E)$ in which the set of vertices V corresponds to unique dk-mers of length smaller or equal to d, and the set of edges E corresponds to unique edges of dk-mers. Two dk-mers have an edge if they are adjacent, sharing $k - 1$ overlap.*

The new algorithm steps are described below:

- Search overlapped regions with length d_1, $k < d_1 < m$, generating one vertex for each unique d_1k-mers and applying the suffix-prefix overlap of $(k - 1)$ length criteria to generate the edges. The result is d_1k-mers vertices and edges sets of extra-compacted *de Bruijn* Graph $G_{d_1,k}$.

– Search overlapped regions with size d_2, $k < d_2 < d_1$, decomposing each vertex in $V_{d_1,k}$ into $d_2 k$-mers. Generate one vertex for each unique $d_2 k$-mers to get a set of vertices and apply the suffix-prefix overlap of $(k-1)$ length criteria to get the set of new edges. The union of new edges and $E_{d_1,k}$ generates $E_{d_2,k}$. The result is $d_2 k$-mers vertices and edges sets of extra-compacted *de Bruijn* Graph $G_{d_2 k}$.

– Search iteratively duplicated regions with size d_i, $k < d_i < d_2 < d_1$, decomposing each vertex in $V_{d_{i-1},k}$ into $d_i k$-mers. Generate one vertex for each unique $d_i k$-mers to get a set of vertices and apply the suffix-prefix overlap of $(k-1)$ length criteria to get the set of edges, adding the edges from $G_{d_{i-1},k}$. The result is $d_i k$-mers vertices and edges sets of extra-compacted *de Bruijn* Graph $G_{d_i,k}$.

– Search for k overlaps at last iteration with $d_z = k$, and $d_z < .. < d_i < ... < d_2 < d_1$, decomposing each vertex in $V_{d_{z-1},k}$ into $d_z k$-mers. Generate one vertex for each unique $d_z k$-mers to get a set of vertices and apply the suffix-prefix overlap of $(k-1)$ length criteria to get the set of edges, adding the edges from $G_{d_{z-1},k}$. Since edges were generated using the suffix-prefix overlap of $(k-1)$ length criteria that appear at least in one read, and $V_{d_{z-1},k}$ corresponds to the set of unique k-mers due to $d_z = k$, the result of this steps is a DBG $G_k(V, E)$ (1) (in Fig. 1b appears a representation of the graph creation overview through a round of iterations).

The update function defines the value of d for each iteration. For instance, we proposed the update function in Eq. 1, in which is used the *step* variable to decrease d in each iteration.

It is worth to note that the suffix-prefix overlap of $(k-1)$ length criteria for edges mentioned above, implies that this overlap exists in at least one read.

$$d_i = update(d_{i-1}, step) = \begin{cases} k & \text{if } d_{i-1} - step < k \\ d_{i-1} - step & \text{otherwise.} \end{cases} \tag{1}$$

4 External Memory Processing at Last Step Analysis

Our approach promotes the idea to process the graph as much as possible in the main memory, reducing the number of duplicated k-mers in each iteration. Only when the available main memory becomes insufficiently to store the structure of $G_{d_i,k}$, the use of an external memory solution is suggested. At that time, large duplicate regions have already been identified. This will allow avoiding to process a significant amount of duplicated k-mers in external memory, and consequently, reducing the number of I/O operations. Moreover, before using external memory processing, it is possible to apply an intermediate tuning solution to reduce even more the amount of data to processes in external memory.

Using our approach is possible to build a macro representation of DBG (i.e., the extra-compacted DBG) in main memory and, if only if necessary, use a solution in external memory in the last iterations. In that sense, our vision is to be able to take better advantage of the available RAM, reaching a higher percentage of processing before going to processing using external memory.

For an iteration i, given $i < z$, in which M is not sufficient to storing $G_{d_i,k}$, it is possible to export $V_{d_{i-1},k}$ and $E_{d_{i-1},k}$, to be used as input of external solutions. The dk-mers into $V_{d_{i-1},k}$ could be exported so that they can be seen as a set of reads R with multiplicities for other solutions. The set of edges $E_{d_{i-1},k}$ is a subset of final E. Then, the set E of external solution could be initialized making $E = E_{d_{i-1},k}$.

The **external memory** model [1], also called the "I/O Model" or the "Disk Access Model" (DAM), is commonly applied in algorithms developed to manage a massive amount of data. It simplifies the memory hierarchy to just two levels. The CPU is connected to a fast cache of size M; this cache, in turn, is connected to a much slower disk of effectively infinite size. Both cache and disk are divided into blocks of size B, so there are M blocks in the cache. Transferring one block from cache to disk (or vice versa) costs 1 unit. Memory operations on blocks resident in the cache are free. Thus, the fundamental goal is to minimize the number of transfers between cache and disk [10].

In that scene, using the case of sorting approach in [11], the collection of all k-mers and edges will be sorted to identify the set of V and E. Therefore the correspondingly optimal number of I/O to get V, given N k-mers is defined by the Equation 2.

$$\Theta\left(\frac{N \log (N/B)}{B \log (M/B)}\right) \quad (2) \quad \Theta\left(\frac{(N - P(i)) \log ((N - P(i))/B)}{B \log (M/B)}\right)$$
$$(3)$$

Definition 4. *A skipped k-mer is a k-mer element that it was not necessary to generate. The $p(i)$ is a number of skipped k-mers for iteration i. The accumulated number of skipped k-mers until iteration z is defined by: $P(z) = \sum_{i=1}^{z} p(i)$*

Using our approach, if there exists an iteration i, given $i < z$, in which M is not sufficient to store $G_{d_i,k}$, it is possible to build the DBG using this sorting approach decreasing the I/O as shown in Eq. 3. In that sense, N is reduced by the number of k-mers that will be avoided to processed $P(i)$.

Since the number of edges to process was reduced as the number of k-mers during previous iterations, the same analysis for I/O could be applied to the set of edges.

Now, we turn to the case of using a partition processing approach in external memory. In that case, we analyzed the I/O in its three main steps: distribution, processing, and merging. In the first step, the collection of all k-mers and edges implicit will be distributed in n partitions. The number of elements and the criteria used to distribution will determine the size and the number of partitions. The distribution of the collections is hard to know until the factual data has been distributed. This fact can cause a re-partition in case the amount of data is more extensive than what is supported to be processed in the main memory. The number of I/O operations in this step is shown in Eq. 4.

In the second step, each partition is read from the disk and processed. Then the results are writing in the disk in a compiled partition. Compiled partitions are less than initial partitions. The I/O of processing step is shown in Equation 5 given $0 < \gamma < 1$. It is used γ to represents the I/O operations consumed during the write of compiled partitions. Finally, the merging step reunites all compiled partitions to generate a result, as is shown in Eq 6. The overall partition strategy I/O is given by Eq 7.

$$\sum_{j=1}^{j=n} \frac{size\_of\_part_j}{B} \quad (4) \qquad (1+\gamma) \sum_{j=1}^{j=n} \frac{size\_of\_part_j}{B} \quad (5)$$

$$\gamma \sum_{j=1}^{j=n} \frac{size\_of\_part_j}{B} \quad (6) \qquad 2(1+\gamma) \sum_{j=1}^{j=n} (\frac{size\_of\_part_j}{B}) \quad (7)$$

Therefore, during the execution of our approach, if there exists an iteration i, given $i < z$, in which M is not sufficient to store $G_{d_i,k}$, it is possible to build the DBG using this distribution approach decreasing the I/O as shown in Eq. 8 given $0 < \gamma < 1$. The number of k-mers impacts in the number and size of partitions, given a reduction of initial k-mers by $P(i)$. To represent this reduction, we used the variable μ, with $\mu > 1$.

Definition 5. μ: *Represents the reduction of initial k-mers by $P(i)$, given $\mu > 1$.*

$$\mathcal{IO} = 2(1+\gamma) \sum_{j=1}^{j=n'} (\frac{size\_of\_part_j}{\mu_j B}) \quad (8)$$

Finally, so that external memory implementations for the construction of DBG can use our approach, we propose that the former implements an input interface, such as:

- The sequence of dk-mers in $V_{d_{i-1},k}$ will treated as reads.
- Initialize the multiplicity for each unique k-mer with the multiplicity of the dk-mer.
- Initialize the set of edges with $E_{d_{i-1},k}$.

5   Implementation and Preliminary Results

In order to validate the approach, an implemented test prototype following the specification exposed in [7] was used. We have run our executions in a virtual machine hosted in private cloud infrastructure, using one virtual machine with Ubuntu 18.04, one CPU core Intel Xeon E312xx 2.2G Hz, with 33 GB of RAM and 500 GB of HD.

The datasets used in our experiments include three groups of organisms:

Sugar Cane Libraries. Fragment libraries collected from Brazilian sugarcane species kept by UFRJ's Institute of Medical Biochemical (IBqM): (i) R03 with $n = 8,520,922$, (ii) R06 with $n = 5,298,464$, both with $m = 72bp$, and (iii) R10 with $n = 5,723,392$ and $m = 76bp$, where n represent the number of reads, and m is the read length.

Human Chromosome 14. Fragment library of Human Chromosome 14 (Ch14) available in http://gage.cbcb.umd.edu/data/: (i) H1: Library 1 with $n = 18,166,705$ and (ii) H2: Library 2 with $n = 18,166,798$, both with $m = 101bp$ in average.

Bombus Impatiens (Bumblebee). Fragment library of Bombus impatiens available in http://gage.cbcb.umd.edu/data/: (i) B2: Library 2 with $m = 124bp$ in average and $n = 120,000,000$.

To validate our experiments, each test was executed in ABYSS to construct the DBG. We have noticed that, for each execution, the number of vertices and edges of DBG produced by ABYSS were equivalent to the DBG output that our approach generates. Table 1 shows our set of planned experiments that helps with the comprehension of our actual contributions.

Table 1. Experiments description. In all cases $step = 10$.

Datasets	k	d_1	Goal
R03, R06, R10, H1, H2	12	64	Test the approach, and proof that it is viable. Measure unique dk-mers
R10	15	52, 55, 58, 61, 64	Shows how the d_1 impact in the number of k-mers skipped from being processed, the accumulated number of processed elements and unique dk-mers
R03, R06, R10, H1, H2	12, 13		Comparing our approach with the requirements for DBG construction of other assemblers like ABYSS and Velvet
Bee	31	100, 55, 35, 31	Proving our approach in case that DBG does not fit in main memory

5.1   Number of Skipped k-mers at Each Iteration

The execution of the first experiment, reveals that our approach reduces the need for processing a significant amount of k-mers. For all datasets, the percent of k-mers that did not have to be processed was over 70%. The accumulate value $P(i)$ means the number of elements that are skipped from being processed. It also shows the remaining elements to process if at that point, the execution needs to be processed in external memory, or even by another approach. The dataset R03 obtains the highest percentage, with a 84,81%, while the highest amount of skipped k-mers was given for the Human Ch14 libraries with more than billion. We may explain this behavior as the number of reads in human libraries is, at least, two times the number of reads when compared to the other datasets.

Analysis of d. How this affects the number of skipped k-mers
To analyze the impact of d value in the number of saved k-mers, were tested five values of d over R10, updating d at each iteration through $update(d) = d_{i-1} - 10$ (Function 1). Depending on the initial d_1, the execution may have more or less number of iterations. As we can see in Fig. 2, all iterations have the same trend over the cumulative percentage of skipped k-mers for different d_i values. The execution that has the greater number of skipped k-mers was the one whose last iteration had a d that eventually came closer to k. The average of the replication factor for execution, meanwhile, showed almost constant behavior overall executions, varying from 1.23 to 1.57, with an average of 1.31.

The observed variation in the number of skipped k-mers demonstrated that it is possible to fine-tune d_1 to obtain betters results.

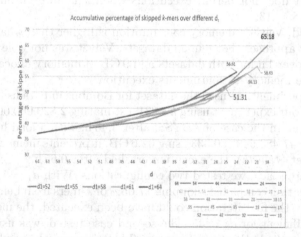

Fig. 2. Comparison of accumulative percentage of skipped k-mers over different executions starting with different d_1.

5.2   Comparison with Other Assemblers

To evaluate the performance of our approach, we compared its results with common assemblers. In that case, we select ABYSS [17] and Velvet [18] as they are commonly used. In competitions such as Assemblathon and Gage they appear as assemblers most frequently used among those selected by competing teams. In both cases, they construct an exact representation of the DBG (in case of ABYSS we executed the version with hash table instead of the BF version).

In the case of ABYSS, it starts generating all k-mers and save the uniques in a hash table using a 2bit codification and a bitmap for edges representation. After getting the set of vertices V, it is traversed, and the edges are generated not over the reads but tested for each k-mer the existence of all possible extensions in V.

Velvet, in turn, has different processing and data structures. Firstly, it generates all k-mers and saves them into a hash table, specifically into a splay tree that resides in each bucket to manage collisions. For each k-mer, the position in the read and the read identifier are tracked, generating the Roadmap file. After, the Roadmap and the sequences files are used as inputs to created the vertices and the edges for complete graph generation.

Using the first experiment as a baseline, we tried to execute the assembly for the same datasets using ABYSS and Velvet with $k = 12$. The DBG construction requirements in ABYSS were comparable with the execution of our implementation for $d_1 = 12$, which is corresponded to the construct directly the DBG, as well as, ABYSS could be compared with the execution of our implementation using our approach through some iterations using $d_1 = 64$ and $step = 10$. Turning to Velvet, it does not permit executions when k is an even number. Thus, we decided use $k = 13$.

As evidenced, Velvet's memory consumption is higher than what is reported when using our approach. For human datasets, Velvet was not able to finish the execution. For over 1 h for both datasets and 9 GB of memory allocated, it could not assign more memory to continue its execution.

As a last experiment, we used the dataset for Bombus impatiens (bumblebee), B2, and we ran our experiment using $k = 31$, estimating 2,820,000,000 k-mers for 30 millions reads. In the case of ABYSS, after 3.45 h, with a load hash factor of $715,400,895/2,147,483,648 = 0.333$ using 32.6 GB, it presents memory insufficient error. For Velvet (*velveth*), the execution crashed after 0.65 h.

Using our approach, we tested two configurations. With $d_1 = 55$ and $step = 10$, the first iteration completed, skipping 17.65% of k-mers from being processed. However, if only the first iteration could have been executed, the memory would have been insufficient for $d_2 = 45$. The second case tested, was using $d_1 = 100$ and using $step = 10$. At this time it was possible to improve the results, obtaining 18.03%, meaning 10,900,042 skipped k-mers more that the previous result. The last iteration completed was $i = 5$, for $d_5 = 60$ using 30.47 GB of memory.

6   Conclusions and Future Works

This paper presents a formal and practical approach that enables an external memory processing to construct a complete and correct *de Bruijn* Graph that does not need to count on all k-mers, as most common approaches do. We have detailed an analytical evaluation that shows that the number of k-mers skipped throughout the execution may considerably impact on the reduction of the number of I/O operations.

Some experiments validate our approach, analyzing the number of skipped k-mers and the way of d_1 affects the execution and make it feasible. We have also considered a comparison with ABYSS and Velvet, two among some common referred and used assemblers. The results show that our approach may be effectively considered for all *De Bruijn* graph based algorithms.

Finally, we have studied a real case in which the internal memory limit is reached and we could skip 18% of k-mers, in this case corresponding to a total of 508,491,754 k-mers, that incurred in a better and more efficient external memory solution.

We are currently interested in the study of the distributions of unique dk-mers to estimate d_1 and *step*, as we have noticed the actual impact of these parameters with respect to the number of skipped k-mers.

References

1. Aggarwal, A., Vitter, J.S.: The input/output complexity of sorting and related problems. Commun. ACM **31**(9), 1116–1127 (1988). https://doi.org/10.1145/48529.48535. https://doi.acm.org/10.1145/48529.48535
2. de Armas, E.M., Ferreira, P.C.G., Haeusler, E.H., de Holanda, M.T., Lifschitz, S.: K-mer mapping and RDBMS indexes. In: Kowada, L., de Oliveira, D. (eds.) BSB 2019. LNCS, vol. 11347, pp. 70–82. Springer, Cham (2020). https://doi.org/10.1007/978-3-030-46417-2_7
3. Bradnam, K.R., et al.: Assemblathon 2: evaluating *de novo* methods of genome assembly in three vertebrate species. GigaScience **2**(1), 1–31 (2013)
4. Chikhi, R., Limasset, A., Jackman, S., Simpson, J.T., Medvedev, P.: On the representation of de Bruijn graphs. In: Sharan, R. (ed.) RECOMB 2014. LNCS, vol. 8394, pp. 35–55. Springer, Cham (2014). https://doi.org/10.1007/978-3-319-05269-4_4
5. Chikhi, R., Limasset, A., Medvedev, P.: Compacting de Bruijn graphs from sequencing data quickly and in low memory. Bioinformatics **32**(12), i201 (2016)
6. Cook, J.J., Zilles, C.: Characterizing and optimizing the memory footprint of de novo short read DNA sequence assembly. In: International Symposium on Performance Analysis of Systems and Software, ISPASS 2009, pp. 143–152 (2009). https://doi.org/10.1109/ISPASS.2009.4919646
7. de Armas, E.M., Castro, L.C., Holanda, M., Lifschitz, S.: A new approach for de bruijn graph construction in *de novo* genome assembling. In: 2019 IEEE International Conference on Bioinformatics and Biomedicine (BIBM), pp. 1842–1849 (2019)
8. de Armas, E.M., Haeusler, E.H., Lifschitz, S., de Holanda, M.T., da Silva, W.M.C., Ferreira, P.C.G.: K-mer Mapping and de Bruijn graphs: the case for velvet fragment assembly. In: 2016 IEEE International Conference on Bioinformatics and Biomedicine (BIBM), pp. 882–889 (2016). https://doi.org/10.1109/BIBM.2016.7822642
9. de Armas, E.M., Silva, M.V.M., Lifschitz, S.: A study of index structures for K-mer mapping. In: Proceedings Satellite Events of the 32nd Brazilian Symposium on Databases. Databases Meet Bioinformatics Workshop, pp. 326–333 (2017)

10. Demaine, E.: Lecture notes in Advanced Data Structures, MIT course number 6.851 (Spring 2012). https://ocw.mit.edu/courses/electrical-engineering-and-computer-science/6-851-advanced-data-structures-spring-2012/calendar-and-notes/MIT6_851S12_L7.pdf

11. Kundeti, V., Rajasekaran, S., Dinh, H.: Efficient parallel and out of core algorithms for constructing large bi-directed de Bruijn graphs. arXiv e-prints (2010)

12. Li, R., et al.: *De novo* assembly of human genomes with massively parallel short read sequencing. Genome Research (2009)

13. Li, Y., Kamousi, P., Han, F., Yang, S., Yan, X., Suri, S.: Memory efficient minimum substring partitioning. Proc. VLDB Endow. **6**(3), 169–180 (2013). https://doi.org/10.14778/2535569.2448951. https://dx.doi.org/10.14778/2535569.2448951

14. Salzberg, S.L., et al.: GAGE: a critical evaluation of genome assemblies and assembly algorithms. Genome Res. **22**(3), 557–567 (2012)

15. Schatz, M.C., Delcher, A.L., Salzberg, S.L.: Assembly of large genomes using second-generation sequencing. Genome Res. **20**(9), 1165–1173 (2010)

16. Silva, M.V.M., de Holanda, M.T., Haeusler, E.H., de Armas, E.M., Lifschitz, S.: VelvetH-DB: data persistency during fragment assembly. In: Proceedings Satellite Events of the 32nd Brazilian Symposium on Databases. Databases Meet Bioinformatics Workshop, pp. 334–341 (2017). (in Portuguese)

17. Simpson, J.T., Wong, K., Jackman, S.D., Schein, J.E., Jones, S.J., Birol, I.: ABySS: a parallel assembler for short read sequence data. Genome Res. **19**(6), 1117–1123 (2009)

18. Zerbino, D.: Velvet software. EMBL-EBI (2016). https://www.ebi.ac.uk/zerbino/velvet/. Accessed 15 June 2019

Computational Methodology for Discovery of Potential Inhibitory Peptides

Vivian Morais Paixão(✉) ⓘ and Raquel C. de Melo Minardi ⓘ

Universidade Federal de Minas Gerais, Belo Horizonte, MG, Brazil
`vivianmp95@ufmg.br`

Abstract. In 2020, a new pandemic caused by a coronavirus has impacted the economic and public health landscape on a global level. Named SARS-CoV-2, it causes COVID-19 and, in two years, has caused thousands of deaths. Among its viral particles, SARS-CoV-2 has an important structural protein called Spike (S), and its entry into human cells is mediated by an interaction between the Spike and the human receptor Angiotensin Converting Enzyme 2 (ACE2). This S/ACE2 binding depends on the cleavage of the Spike into three parts (S1, S2 and S2') by host cell proteases. For this, the S protein undergoes a conformational change that exposes a cleavage site between the S1 and S2 domains, being initially cleaved by the Furin enzyme. The S2 part is cleaved by TMPRSS2 (Transmembrane Serine Protease II) to expose the fusion peptide, promoting endocytic entry of the virus. TMPRSS2 can be inhibited by clinically approved serine protease inhibitors, making it a promising target for the treatment of viral infections. Consequently, our objective was to look for peptides that weren't described as inhibitors for SARS-CoV-2 but can be repositioned. In this paper, we propose a computational method to collect, filter, simulate protein-peptide interaction and identify the best hits based on the pattern of interactions. In addition to the main contribution of the paper that is the method, another contribution of this work is the proposal of candidate peptides.

Keywords: SARS-CoV-2 · TMPRSS2 · Peptides

1 Introduction

Coronaviruses are positive-sense single-stranded RNA (ssRNA+) viruses that cause respiratory infections in a variety of animals. In 2020, a new pandemic caused by a coronavirus, named SARS-CoV-2, generated economic and public health impacts at a global level. According to studies, the severity of symptoms ranges from mild to critical, with a fatality rate among critical cases of around 49% [1].

SARS-CoV-2 particles contain four primary structural proteins: spike (S), membrane (M), envelope (E) and nucleocapsid (N) proteins [1]. Virus entry into human cells is mediated by an interaction between S-glycoprotein and the Angiotensin Converting Enzyme 2 (ACE2) receptor. A key point is that binding of S to ACE2 depends on cleavage of the protein into three parts (S1, S2 and S2') by host cell proteases, typically

© The Author(s), under exclusive license to Springer Nature Switzerland AG 2022
N. M. Scherer and R. C. de Melo-Minardi (Eds.): BSB 2022, LNBI 13523, pp. 91–96, 2022.
https://doi.org/10.1007/978-3-031-21175-1_10

by transmembrane serine protease 2 (TMPRSS2). For virus entry, the S protein undergoes a conformational change that exposes a cleavage site between the S1 and S2 domains, being initially cleaved by the Furin enzyme. The S2 part is cleaved by TMPRSS2 to expose the fusion peptide, promoting endocytic entry of the virus [3].

TMPRSS2, the target of our study, is a transmembrane serine protease that has a trypsin-like C-terminal domain with a canonical catalytic triad Ser441-His296-Asp345 [4]. It belongs to the S1 family of serine proteases with cleavage activity at Arg or Lys residues [5–7] and, overall, its active can bind to various substrate sequences with the strictest preference for positions P1 and P2 [4].

This molecule can be inhibited by clinically approved serine protease inhibitors, making it a promising target for the treatment of viral infections. Consequently, our objective was to look for peptides that were not described as inhibitors for SARS-CoV-2 but can be repositioned. Furthermore, a unique benefit of blocking TMPRSS2 and related airway proteases is that in addition to the coronavirus, several other respiratory viruses can be targeted, such as the influenza virus. In this article, we propose a computational method to collect, filter, simulate the protein-peptide interaction, and thus identify the best hits based on the pattern of interactions, acquiring candidates for inhibitors of a particular molecule to be studied. In addition to the main contribution of the work, which is the method, we proposed candidate peptides to inhibit the protease of our case study.

2   Methodology

2.1   Bibliographic Survey

Through a bibliographic survey, information was acquired about the catalytic site of TMPRSS2 and its mechanism of action in other molecules. We obtained information that its active site has three chains, two of which are non-catalytic (LDLR and SRCR) and the catalytic chain SP (serine peptidase), with a canonical catalytic triad Ser441-His296-Asp345. In addition, this site has a binding preference for the P1 and P2 positions [4] and it is an Arg/Lys-specific protease.

2.2   Catalytic Site Analysis

Using the information obtained in the previous step, we analyzed one of the TMPRSS2 structures available in the Protein Data Bank (PDB ID: 7MEQ; resolution 1.95 Å; Fraser et. al., April 2021) and selected the catalytic site (Ser441-His296-Asp345) using the PyMOL software (The PyMOL Molecular Graphics System, Version 1.2r3pre, Schrödinger, LLC.) and Fig. 1, below, was generated.

2.3   Catalytic Site Analysis

From the structure, we took the residues that were up to 5 angstroms away from the catalytic site and then submitted the 7MEQ structure and the selected residues to the Propedia tool [http://bioinfo.dcc.ufmg.br/propedia/]. Propedia is a database that permits

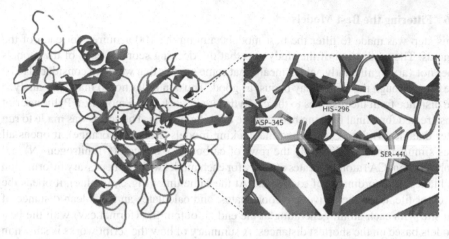

Fig. 1. Structure of the serine protease TMPRSS2 (PDB ID: 7MEQ).

clustering, searching, and visualizing of protein-peptide complexes according to varied criteria. Adding these residues up to 5 angstroms from catalytic site as criteria, this tool returned 127 peptides that have the pattern of binding to this site. A crucial point is that we know that the resulting peptides are candidate inhibitors and are not cleaved by the enzyme because the peptides selected through Propedia are already found in crystallographic files, so they were not cleaved by the enzyme.

2.4 Sequence Filtering

To filter the best sequences, some criteria were defined: (I) have at least 8 residues, (II) have arginine or lysine (due to protease preference) and (III) have no arginine at the end of the sequence (may not be an entire peptide). This initial filtering resulted in 37 sequences. Another criterion studied was the correspondence of the sequence with the sequence of aprotinin, a known serine protease inhibitor, through a pair-wise alignment with each of the peptides. This filtering was done by taking those alignments that had at least 3 matches with the sequence of the aprotinin loop that binds to TMPRSS2 and results in its inhibition [8]. This second filtering resulted in 11 sequences.

2.5 Molecular Docking

For the binding analysis of each of the 11 selected peptides with TMPRSS2, the 7MEQ structure and the 11 sequences were subjected to molecular docking by HPEP-DOCK [http://huanglab.phys.hust.edu.cn/hpepdock/] (the two smallest sequences), and by HDOCK [http://hdock.phys.hust.edu.cn/], with the inclusion of the protease active site as a parameter. 100 docking models were generated for each of the sequences, obtaining a total of 1100 files.

2.6 Filtering the Best Models

This step was made to filter the best models, among the 100 acquired for each of the ligands. The motivation of the study was that the docking score of each of the models does not vary significantly, which means that some models that were not considered to be the best, through the score, may be just as good. Thus, in order not to manually analyze the distance from the residues to the catalytic site of each of the models, a Python script was created for a final filtering between the docking models. The script was made to run after data acquisition. Concerning the docking models (Fig. 2, in orange), it opens all files simultaneously, takes only the rows of carbon (C), oxygen (O), nitrogen (N) and alpha carbon (CA) atoms, creates an entry for each model with the necessary information and takes the coordinates of each atom. At the same time (Fig. 2, in blue), it opens the receptor file, takes the active site coordinates, and calculates an Euclidean distance of the site with each atom, generating at the end an output file (format.csv) with the best models based on the shortest distances. A summary of how the script works is shown in Fig. 2, below:

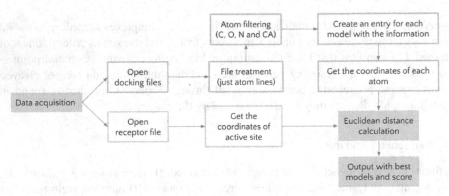

Fig. 2. Scheme showing how the Python script works to filter the generated docking models. (Color figure online)

Based on this methodology and the return of the best docking models between the receptor and each ligand, a visual analysis was performed using PyMOL. The chosen ligands can be seen in Table 1 and Fig. 3 (results).

3 Results

Our script returned the 10 best models (out of 100) of each of the 11 peptides based on the criterion of their atoms being less than 5 angstroms away from the active site. We analyzed these models using PyMOL Software and selected those who had a loop containing arginine or lysine in a short distance from the active site. The results are seen in Table 1, where the six chosen peptides are represented with their respective PDB ID, name, best structural model and remark score (represents the quality of the model, according to the HDOCK website). The models can be seen anchored to the protease in Fig. 3. An important observation is that two of the chosen ones are the same peptide (Antileukoproteinase) anchored to different structures (PDB ID: 2Z7F, PDB ID: 4DOQ).

Table 1. Table containing the best inhibitors select using this project methodology.

ID	Peptide	Best model	Remark score
1AN1	Tryptase inhibitor (*Hirudo medicinalis*)	98	−133.14
1GL1	Protease Inhibitor LCMI II (*Locusta migratoria*)	81	−68.27
2Z7F	Antileukoproteinase (*Homo sapiens*)	23	−128.87
3FP7	Pancreatic trypsin inhibitor	81	−105.17
4DOQ	Antileukoproteinase (*Homo sapiens*)	23	−128.87
4ZKN	Upain-1-W3A (murinised human uPA I Hydrolase inhibitor)	64	−125.454

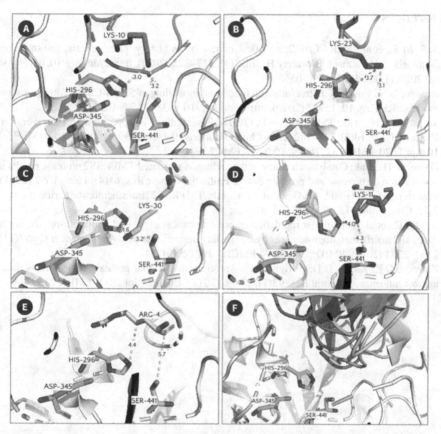

Fig. 3. Best models anchored to the catalytic site of TMPRSS2 (His296, Asp345 and Ser441): (A) PDB ID: 1AN1 (shown as pink), peptide: chain B; (B) PDB ID: 1GL1 (shown as purple), peptide: chains D, E, F; (C) PDB ID: 2Z7F (shown as green), peptide: chain B; (D) PDB ID: 3FP7 (shown as pink), peptide: chain B; (E) PDB ID: 4ZKN (shown as yellow), peptide: chain B; and (F) all peptides overlapped and represented as cartoons. (Color figure online)

4   Conclusion

In conclusion, this methodology is able to give visibility to peptides that initially would not be considered for evaluation, since their scores were slightly under the best 10 models, but they proved to have a better fitting when visualizing the whole structure. In addition, five peptides were selected as potential candidates for TMPRSS2 inhibitors.

As next steps, we are planning to add negative controls (decoys) in our experiment to show the difference between peptides that bind to the receptor from peptides that don't bind and, going forward, we intend to perform in vitro experiments to analyze the results obtained in silico.

References

1. Sofi, M.S., et al.: SARS-CoV-2: a critical review of its history, pathogenesis, transmission, diagnosis and treatment. Biosafety Health **2**(4), 217–225 (2020). https://doi.org/10.1016/j.bsh eal.2020.11.002. ISSN 2590-0536
2. Brooke, G.N., Prischi, F.: Structural and functional modelling of SARS-CoV-2 entry in animal models. Sci. Rep. **10**, 15917 (2020). https://doi.org/10.1038/s41598-020-72528-z
3. Stopsack, K.H., et al.: TMPRSS2 and COVID-19: serendipity or opportunity for intervention? Cancer Discov. **10**(6), 779–782 (2020). https://doi.org/10.1158/2159-8290.CD-20-0451. Epub 10 Apr 2020. PMID: 32276929; PMCID: PMC7437472
4. Daniel, E.H., et al.: Catalytic cleavage of the androgen-regulated TMPRSS2 protease results in its secretion by prostate and prostate cancer Epithelia. Cancer Res. **61**(4), 1686–1692 (2001)
5. Thunders, M., Delahunt, B.: Gene of the month: TMPRSS2 (transmembrane serine protease 2). J. Clin. Pathol. **73**, 773–776 (2020)
6. Reid, J.C., et al.: Pericellular regulation of prostate cancer expressed kallikrein-related peptidases and matrix metalloproteinases by cell surface serine proteases. Am. J. Cancer Res. **7**(11), 2257–2274 (2017). PMID: 29218249; PMCID: PMC5714754
7. Zhirnov, O.P., Klenk, H.D., Wright, P.F.: Aprotinin and similar protease inhibitors as drugs against influenza. Antiviral Res. **92**(1), 27–36 (2011). Epub 23 Jul 2011. PMID: 21802447

A Non Exhaustive Search
of Exhaustiveness

Letícia Kristian Silva Cecotti⬛, Maurício Dorneles Caldeira Balboni⬛,
Oscar Emilio Arrúa Arce⬛, Karina dos Santos Machado$^{(\boxtimes)}$⬛,
and Adriano Velasque Werhli⬛

Computational Biology Laboratory - COMBI-LAB, Centro de Ciências
Computacionais, Universidade Federal do Rio Grande - FURG, Av. Itália, km 8.,
Rio Grande, RS, Brazil
{balboni,oarrua88,karina.machado,werhli}@furg.br

Abstract. Knowledge of how small molecules interact with target proteins is of great interest in many fields, especially in drug discovery. Generally, this knowledge is obtained in time-consuming and very expensive wet experiments, which emphasises the importance of *in silico* computational predictions by docking simulations. There are many available docking software and among them Autodock Vina is one of the most accurate and largely applied in many studies. In Autodock Vina, among the different parameter settings, the Exhaustiveness is of crucial importance as it is directly related to the accuracy of the resulting poses. In this work, we investigate the Exhaustiveness parameter in a set of 4,463 protein-ligand complexes (PDBbind2018 refined dataset) for which the correct ligand pose is known. The quality of the Autodock Vina results is assessed by the distance between the experimental and the predicted ligand poses and by the Free Energy of Binding calculated by Autodock Vina. The main purpose of the analysis discussed in this paper is to help users define the Exhaustiveness parameter and thus achieve a good trade-off between simulation time and pose prediction quality. The results suggest that there is no difference whether several simulations with small Exhaustiveness or a few simulations with high Exhaustiveness setting are performed. In addition, the results give a good indication of the number of simulations and Exhaustiveness setting required for meaningful docking results with AutoDock Vina.

Keywords: Molecular docking · Parameter setting · Autodock Vina

1 Introduction

The achievements in bioinformatics and computational biology in recent decades are remarkable. They are the result of an interplay of several factors, e.g., the

This study was supported by CAPES Edital Biologia Computacional (51/2013), CAPES Financial Code 001 and CNPq (439582/2018-0).

N. M. Scherer and R. C. de Melo-Minardi (Eds.): BSB 2022, LNBI 13523, pp. 97–108, 2022.
https://doi.org/10.1007/978-3-031-21175-1_11

tremendous developments in computer hardware and developments in biochemistry, as well as mathematical, statistical, and computational methods. One of the most successful computational methods in these fields is molecular docking, hereafter referred to simply as docking. Docking is a computational method (*in-silico* method) that is routinely used in biology laboratories around the world for drug screening, protein-protein interactions, and nanomaterial behavior studies, among other applications [20].

Docking is generally understood to be the computational process of estimating the affinities and pose of small molecules when they bind to a macromolecule, protein, or enzyme [12]. Docking experiments have yielded very good results in many different studies. Interestingly, the computational resources required are usually small, which is why docking is ubiquitous as a standard procedure prior to experiments in the wet lab.

There are dozens of protein ligand software packages available. The website [1] provides a good set of software and web servers for docking. In addition, some reviews present the most common docking software packages from different points of view, e.g., Pagadala et al. discuss [16] on rigid and flexible docking, Su et al. [19] presents a comparative analysis of 25 scoring functions (scoring benchmark), and Garcia-Godoy et al. [8] address single- and multi-objective meta-heuristics for molecular docking optimization problems. Among all docking programmes presented in these reviews, we can highlight as examples of docking programmes: AutoDock Vina [22], Glide [7], GOLD [11], GlamDock [21], etc.

Among all these options, Autodock Vina is one of the most widely used software for protein-ligand docking [2]. The reasons for its popularity are the quality of results, speed, flexibility in configuration, and the fact that it has a free academic license. Moreover, Vina scoring showed good results in scoring and screening power and the best results in docking power in CASF-2016 [19]. In addition, the scoring function Δ_{Vina} RF $_{20}$ [23], which proposes a parametrization framework combining the Vina scoring function with Random Forest models, achieves the best ranking power and scoring power and the second best docking and screening power [19] in CASF-2016. Since Vina originates from the same laboratory as Autodock, it is well known and usually used as the first option in docking studies, which is why it is cited in many scientific publications.

The most recent version of AutoDock Vina, 1.2.0, features new implementation of docking methods, a new mode to fit large number of ligands and an easier way of implementing python scripts [5].

Despite its popularity and quality, the parametrization of docking in Vina deserves further investigation. In particular, it is known that the setting of Exhaustiveness has a large impact on the quality of docking results. Exhaustiveness is a parameter that automatically sets all the parameters of the evolutionary algorithm that Vina applies when searching for the best pose. The default Exhaustiveness in Vina is 8, increasing this value gives a more consistent docking result [15]. Therefore, the main goal of this paper is to investigate how the Vina parameter Exhaustiveness affects the quality of the results. In addition, two different simulation scenarios are investigated: (i) few simulation runs and high Exhaustiveness versus (ii) many simulation runs and low Exhaustiveness.

2 Related Work

The prediction of non-covalent binding between a receptor and a ligand *in silico* is known as molecular docking. The main goal of molecular docking is to predict the pose and binding affinity of these molecules [22].

In [9] it was shown that the time Vina takes to run a simulation does not depend on the search space volume. That is, the size of the docking box does not affect the number of runs each simulation performs. The same study also shows that simulation time varies linearly with the value of Exhaustiveness, as explained in the Vina manual. Given this behavior, it is expected that for the same Exhaustiveness, the probability of finding the correct answer is lower when the search volume is larger.

In [17], the authors discussed that the instructions in the Vina manual are not sufficient to inform the setting of Exhaustiveness and to ensure reproducibility of results in the context of peptide docking. However, such an investigation in the context of small ligands is still pending.

In Devaurs et al. [3], the authors analyzed the effects of distributed and incremental approaches on the accuracy and performance of docking large ligands (especially peptides). They focused on protein-ligand complexes that could not be accurately docked using classical tools and compared the results of enhanced conformational selection by Vina (the Exhaustiveness parameter), running multiple short instances of Vina in parallel and grouping their results (a protocol defined as Multi-Vina), and using a distributed incremental meta-docking method called DINC [4]).

An interesting question raised in [9] is whether running one simulation with high Exhaustiveness or multiple simulations with low Exhaustiveness has an impact on the ability to find the minimum binding energy. The time required in both scenarios is similar. However, in the first scenario, all simulations have the same simulation initialization (seed), while in the second scenario, a different simulation seed is used for each simulation, potentially leading to a greater variety of responses and thus increasing the chance of finding the minimum binding energy. Therefore, in the present work, we perform a series of simulations to better understand which of the previously presented scenarios is best in terms of finding the correct response, i.e., the best pose.

3 Data and Simulations

3.1 Data

In this study, the complexes (protein+ligand) are selected from the PDBbind database [14]. The compilation of the dataset is based on protein-ligand complexes with high-quality crystal structures and their reliable binding data. PDB-bind provides information on various biomolecular complexes along with their experimental binding data [13,24]. The PDBbind database provides a subset of the main base of biomolecular complexes called the "refined set". In the refined set, complexes are selected according to rules related to the quality of the complex structure, the quality of the binding data, and the type of complex. Examples of each of these rules: (i) only non-covalent complexes with a structural resolution of less than 2.5 and an R-factor of less than 2.5 are accepted; (ii) complexes with extremely low (K_d or $K_i > 10$ mM) or extremely high (K_d or $K_i < 1$ pM) binding affinities are not selected; and (iii) only ligand molecules composed of specific atoms (carbon, nitrogen, oxygen, phosphorus, sulphur, halogen, and hydrogen) are selected [14].

In this work, we use the PDBbind refined set 2018 which consists of 4,463 protein-ligand complexes (first step in the methodology flowchart described in Fig. 1).

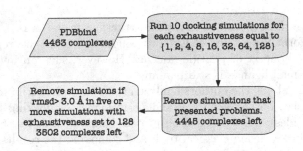

Fig. 1. Flowchart showing data set pre-processing for analysis.

3.2   Simulations

In this study, we want to investigate how results of Vina are affected by the choice of Exhaustiveness. In the present case, we have access to the experimental (true) pose of the ligand and therefore know whether or not a docking simulation finds the correct answer, or at least a good one. Two measures that are generally available to evaluate a docking simulation are the Free Energy of Binding (FEB) and the Root Mean Square Deviation (RMSD). In Vina, FEB is the value obtained from the implemented score function. RMSD in this work is defined as the average distance between the atoms of two ligands, one from PDBBind and the other from the Vina result.

Normally, the user only has access to FEB values obtained in a docking simulation. Therefore, we want to investigate the relationship between FEB and RMSD to understand to what extent the quality of docking results can be judged from FEB values alone.

We run simulations for a set of eight values of Exhaustiveness, $E = \{1, 2, 4, 8, 16, 32, 64, 128\}$. For each of these values, we repeat the simulations 10 times with different initializations (seeds) totalizing 80 simulations for each

PDBBind complex. Since we have 4, 463 complexes, the total number of docking simulations is 357, 040.

3.3   Exhaustiveness

Studies show that the execution time of Vina is linearly proportional to the Exhaustiveness parameter [10]. According to the Vina manual, increasing Exhaustiveness leads to an exponential decrease in the probability of not finding the minimum FEB. This is because increasing Exhaustiveness causes the search algorithm to run more times in each simulation. On the other hand, setting Exhaustiveness low can save time and most likely leads to premature results. Investigating the balance between the execution time and the quality of the results with respect to the Exhaustiveness setting can contribute to a better use of computational resources while maintaining the quality of the results.

In general, the minimum value of Exhaustiveness that is accepted as sufficient to find meaningful results is 128 [6]. However, it is known that in some cases for Exhaustiveness values below than 128 some results are as good as those obtained with an Exhaustiveness value of 128 or higher.

3.4   Simulation Box

Docking box definition is a crucial step in docking simulation with Vina. Docking boxes were automatically defined using the pocket information from the PDBind files. The pocket information is used to check which of the protein chains interact with the ligand. Only the interacting chains are kept in the protein files prepared for docking. Finally, the docking box is defined as the one that includes all the selected chains.

Preparation of pdb files, conversion to pdbqt format, and selection of interacting chains is done using scripts written in Python and in Pymol [18]. The ligands are also converted to the pdbqt file type. In addition, all ligands were considered rigid in all simulations.

4   Results

After running the docking simulations as previously explained, 5 complexes were found to have incorrect atomic syntax, 3 complexes had a problem with insufficient memory, and 10 complexes had an unknown error. Thus, we ended up with 4445 biomolecular complexes for which the entire simulation could be completed without any problems.

Since the main interest here is to study the influence of Exhaustiveness on the quality of the results, we decided to remove cases where Vina is unable to find the correct pose. These may include cases where Vina's scoring function is unable to find the correct pose of the ligand, or where the searching boxes were poorly defined. Therefore, we removed all complexes for which more than 50% (5 or more out of 10) of the simulations with Exhaustiveness set to 128 did not achieve RMSD $\leq$ 3 Å. Considering this rule, 943 complexes were removed from the data set before further analysis was performed. The final data set contains 3, 502 complexes.

The results of all simulations are presented in this section. Figure 2 shows a boxplot of FEB for each of the Exhaustiveness settings. Each box of the boxplot represents 35, 020 simulations that are the result of 3, 502 different protein-ligand dockings, each repeated 10 times. Median, first and third quantiles are shown as horizontal lines and outliers as dots.

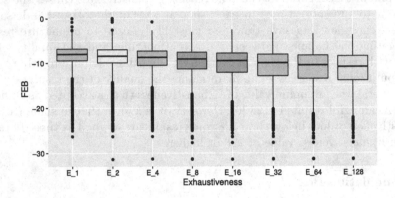

Fig. 2. Boxplot of the FEB for each exhaustiveness.

Figure 3 shows the RMSD distribution for each Exhaustiveness setting in a boxplot. The number of simulations and the meaning of the graphical elements are the same as for the FEB boxplot.

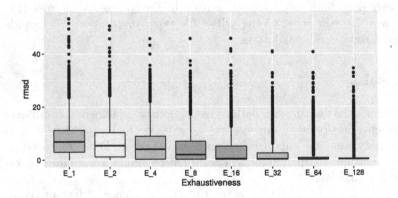

Fig. 3. Box plot of the RMSD for each exhaustiveness.

One of the most important questions when running docking simulations with Vina is whether the search algorithm found a meaningful result. Usually, lower

FEB values indicate better results from the algorithm, but it may be that the low value is related to a local minimum rather than indicating a really good pose of the ligand. In this study, we have proposed a simulation setting where the RMSD is available to us and thus we can determine whether the simulations have converged to a good result or not. However, it is important to emphasise that the RMSD value cannot be used to evaluate docking simulations with new protein-ligand complexes. Consequently, the quality of the results must be evaluated using the FEB value or other clear indicators of the complex under study. Therefore, we created an analysis shown in Fig. 4 with the intention of revealing the behaviour of the Vina search algorithm given the different Exhaustiveness.

To produce the Fig. 4 we have assumed: (i) a simulation has found the correct pose if the resulting RMSD is lower or equal to 3 Å; (ii) different number of simulations are considered: sim_size = {1, 2, 4, 6, 8, 10}. Having established these conditions, for each set of simulations, we select the one that has the lowest FEB and verify, using its RMSD whether it has converged or not. Using the number of converged simulations, we calculate the percentage of the "correct" simulations. The resulting percentage of correct simulations is listed in Table 1. Note that only the FEB values are evaluated to find the best simulation to be selected. This is to mimic the real world, where only the FEB values are available. In fact we are counting how many times the correct pose will be chosen if the lowest FEB simulation is selected from a set of simulations.

In Fig. 4, the horizontal axis shows the different Exhaustiveness and the vertical axis shows the percentage of docking simulations that have converged, i.e., are "correct"'. Each line represents a different sized set of considered simulations, either 1, 2, 4, 6, 8 or 10. For example, considering Exhaustiveness 1, if only one docking simulation is performed the percentage of correct answers is 25.84% (red line in Fig. 4). However, if ten simulations are carried out, using the same setting as before, 60.11% of the answers are correct (purple line in Fig. 4).

Another interesting question in running docking simulations with Vina is whether it is best to run several small simulations (low Exhaustiveness) or a few long simulations (high Exhaustiveness). To try to address this question, Table 2 is presented. This table lists only the number of simulations that are multiples of 2 and whose execution time can therefore be compared to the Exhaustiveness setting. Each coloured diagonal shows simulations and Exhaustiveness that have equivalent execution times. That is, running 2 simulations with Exhaustiveness set to 2 is approximately equivalent to running 4 simulations with Exhaustiveness set to 1.

5 Discussion

The main objective of the present study is to find out how to properly adjust the Exhaustiveness in Vina in order to optimize the reliability of the results. To this end, a series of docking simulations are performed with complexes that we know Vina can find reasonable poses. These complexes are selected based on their RMSD from the docking simulations. For some of the complexes, we

Table 1. Present the percentage of correct results found when considering different number of simulations. The number of simulations are represented in lines and the columns are related with the Exhaustiveness setting. One simulation is correct if its RMSD ≤ 3 Å.

Sim size	% Correct by exhaustiveness							
	E_1	E_2	E_4	E_8	E_16	E_32	E_64	E_128
1	25.84	33.07	44.66	56.65	67.27	78.10	86.55	94.46
2	34.81	43.66	55.08	67.50	78.18	86.69	93.06	97.87
4	44.94	54.94	67.07	78.21	87.01	93.27	96.72	99.17
6	51.37	62.51	73.16	83.01	91.49	96.29	98.06	99.29
8	56.14	67.10	77.01	86.55	93.71	97.28	98.48	99.29
10	60.11	70.76	80.15	88.83	95.11	97.74	98.71	99.29

Table 2. In this table we keep only the results where the number of simulations are multiple of 2 and therefore can have their execution time compared with the Exhaustiveness setting. Each coloured diagonal presents simulations and Exhaustiveness that have equivalent execution times. That is, to run 2 simulations with Exhaustiveness equals 2 is the equivalent of running 4 simulations with Exhaustiveness equals 1. Interestingly, these results show that there is no clear advantage in choosing to run more simulations with lower Exhaustiveness or fewer simulations with higher Exhaustiveness as the number of correct simulations are equivalent in both cases.

Sim size	E_1	E_2	E_4	E_8	E_16	E_32	E_64	E_128
1	25.84	33.07	44.66	56.65	67.27	78.10	86.55	94.46
2	34.81	43.66	55.08	67.50	78.18	86.69	93.06	97.87
4	44.94	54.94	67.07	78.21	87.01	93.27	96.72	99.17
8	56.14	67.10	77.01	86.55	93.71	97.28	98.48	99.29

never obtained good RMSD values, indicating that either the Vina function was not working properly, the search boxes were poorly defined, or there were other unknown reasons. These complexes were excluded because they likely undermine the present study.

A summary of the results are presented in Figs. 2 and 3 in the form of box-plots. As expected, the FEB decreases as the Exhaustiveness is incremented. This indicates that, in many simulations with low Exhaustiveness, the search algorithm has not found the correct, or near correct, answer. This is confirmed by the box plot of RMSD, in Fig. 3, that shows a steady decrease of its values as the Exhaustiveness increases.

A careful inspection of Fig. 2 shows that for Exhaustiveness greater than 4 there are no outliers above the median and the number of outliers below the median increases for the Exhaustiveness greater than 16. It is also observed that the number of cases falling between the first and third quartiles increases with

Exhaustiveness, while the number of outliers decreases, suggesting that a larger Exhaustiveness leads to a lower variance in the results.

Interestingly, Fig. 3 clearly shows that even when the Exhaustiveness is set to 128, there are still outliers, indicating that some simulations have not converged. Thus, we can conclude that setting Exhaustiveness to 128 is not sufficient to guarantee that all simulations converge to acceptable responses.

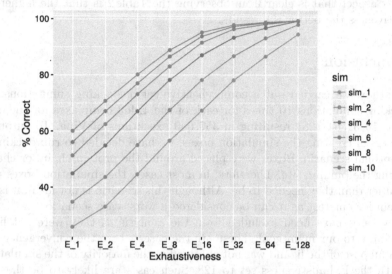

Fig. 4. In this graph, five distinct scenarios are explored. Each coloured line represents a different number of simulations considered, namely 1, 2, 4, 6, 8 and 10. The horizontal axis represents the Exhaustiveness setting. The vertical axis presents the percentage of simulations that are found to be correct. A simulation is considered to be correct if the RMSD is lower or equal to 3 Å. The pose chosen to have its RMSD calculated is the one with the lowest FEB. For example, the red line represents the case where only 1 simulation is run. (Color figure online)

Figure 4 undoubtedly shows that the number of correct predictions increases monotonically with the Exhaustiveness setting, independent of the number of simulations considered. Moreover, an increase in the number of simulations increments the total correct predictions. Considering Exhaustiveness 1, if we run 1 simulation we find only 25.84% of correct simulations, however, if we run 10 simulations the number o correct simulations raises to 60.11%. If we run only one simulation and change the Exhaustiveness to 128, we achieve a 94.46% of correct simulations.

In the Table 2 only the number of simulations and the Exhaustiveness that are multiple are presented. It is known that the time a docking simulation takes is linearly correlated with the Exhaustiveness, thus, a simulation with Exhaustiveness set to 2 takes twice the time of a simulation with Exhaustiveness set to 1. The table shows clearly that for the same equivalent time, the percentage of correct answers is the same. This means that there is no advantage in running many simulations with small Exhaustiveness over running few simulations with high Exhaustiveness.

For example, one can run 8 simulations with Exhaustiveness set to 16, or 1 simulation with Exhaustiveness set to 128 and the number of correct answers will be approximately the same, 93.71% vs 94.46%. In practice, it is easier to run 1 simulation instead of running 8 simulations.

Other aspect that is clear from observing the Table 2 is that the higher the Exhaustiveness the better the results.

6   Conclusion

In this study, we carried out a comprehensive set of docking simulations. We ran docking simulations 10 times for each of the 4, 463 complexes for the $ex = \{1, 2, 4, 8, 16, 32, 64, 128\}$ resulting in 357, 040 executions of Vina. To automate the process of creating the simulation boxes, we have decided to run docking in a non-optimal scenario. Boxes were placed around the protein chain, or chains, containing the binding site. Therefore, in most cases, the simulation boxes were much larger than they needed to be. Although this scenario is not ideal, it is not a problem for our test as it can be considered a worst case scenario.

Many cases have been excluded from the analysis as they were not likely to contribute to our main goal. These are the cases where the convergence to a meaningful pose of the ligand was not obtained in the majority of the simulations even with the Exhaustiveness set to 128. Such cases are likely to be the ones where Vina's scoring function does not perform well.

The results of the simulations show undoubtedly that higher Exhaustiveness leads to better results. Also, it shows that even with the Exhaustiveness of 128 and 10 simulation repeats, some results have not converged to a good ligand pose.

Interestingly, Table 2 exemplifies that there is no "*free lunch*" when it comes to trading the number of simulations for Exhaustiveness. Clearly, the different seeds, i.e. initialization, in the various simulations scenario, is ineffective in better sampling the search space when compared with the few long simulations scheme. Approximately the same percentage of correct results is only achieved if the same computational time is expended, it does not make a difference whether the time is consumed in a few long simulations or many short ones. However, in practise, it is easier to run few simulations than many, thus, one outcome is that few simulations with high Exhaustiveness should be favoured.

The main conclusion of the present study is that the higher the Exhaustiveness the better. Moreover, it does not make sense to set up and run many small simulations because few long simulations are very likely to produce equally accurate results. Also, it is suggested that the docking set with an Exhaustiveness of 128 and with four (4) repetitions, or more, will find the correct pose in more than 99% of the cases.

References

1. Click2drug homepage. http://www.click2drug.org/. Accessed 18 July 2022
2. Chen, Y.C.: Beware of docking! Trends Pharmacol. Sci. **36**(2), 78–95 (2015)
3. Devaurs, D., et al.: Using parallelized incremental meta-docking can solve the conformational sampling issue when docking large ligands to proteins. BMC Mol. Cell Biol. **20**(1), 1–15 (2019)
4. Dhanik, A., McMurray, J.S., Kavraki, L.E.: DINC: a new AutoDock-based protocol for docking large ligands. BMC Struct. Biol. **13**(1), S11 (2013). https://doi.org/10.1186/1472-6807-13-S1-S11
5. Eberhardt, J., Santos-Martins, D., Tillack, A.F., Forli, S.: AutoDock vina 12 0: new docking methods, expanded force field, and python bindings. J. Chem. Inf. Model. **61**(8), 3891–3898 (2021)
6. Forli, S., Huey, R., Pique, M.E., Sanner, M.F., Goodsell, D.S., Olson, A.J.: Computational protein-ligand docking and virtual drug screening with the AutoDock suite. Nat. Protoc. **11**(5), 905–19 (2016)
7. Friesner, R.A., et al.: Extra precision glide: docking and scoring incorporating a model of hydrophobic enclosure for protein-ligand complexes. J. Med. Chem. **49**(21), 6177–6196 (2006). https://doi.org/10.1021/jm051256o. PMID: 17034125
8. García-Godoy, M.J., López-Camacho, E., García-Nieto, J., Del Ser, J., Nebro, A.J., Aldana-Montes, J.F.: Bio-inspired optimization for the molecular docking problem: state of the art, recent results and perspectives. Appl. Soft Comput. **79**, 30–45 (2019)
9. Jaghoori, M.M., Bleijlevens, B., Olabarriaga, S.D.: 1001 ways to run AutoDock vina for virtual screening. J. Comput. Aided Mol. Des. **30**(3), 237–249 (2016). https://doi.org/10.1007/s10822-016-9900-9
10. Jaghoori, M.M., Van Altena, A.J., Bleijlevens, B., Olabarriaga, S.D.: A grid-enabled virtual screening gateway. In: 2014 6th International Workshop on Science Gateways, pp. 24–29. IEEE (2014)
11. Jones, G., Willett, P., Glen, R.C., Leach, A.R., Taylor, R.: Development and validation of a genetic algorithm for flexible docking. J. Mol. Biol. **267**(3), 727–748 (1997)
12. Keretsu, S., Bhujbal, S.P., Cho, S.J.: Rational approach toward COVID-19 main protease inhibitors via molecular docking, molecular dynamics simulation and free energy calculation. Sci. Rep. **10**(1), 1–14 (2020)
13. Liu, Z., et al.: PDB-wide collection of binding data: current status of the PDBbind database. Bioinformatics **31**(3), 405–412 (2014)
14. Liu, Z., et al.: Forging the basis for developing protein-ligand interaction scoring functions. Acc. Chem. Res. **50**(2), 302–309 (2017)
15. Nguyen, N.T., et al.: AutoDock vina adopts more accurate binding poses but AutoDock4 forms better binding affinity. J. Chem. Inf. Model. **60**(1), 204–211 (2019)
16. Pagadala, N.S., Syed, K., Tuszynski, J.: Software for molecular docking: a review. Biophys. Rev. **9**(2), 91–102 (2017). https://doi.org/10.1007/s12551-016-0247-1
17. Rentzsch, R., Renard, B.Y.: Docking small peptides remains a great challenge: an assessment using AutoDock vina. Brief. Bioinform. **16**(6), 1045–1056 (2015)
18. Schrödinger, LLC: The PyMOL molecular graphics system, version 1.8 (2015)
19. Su, M., et al.: Comparative assessment of scoring functions: the CASF-2016 update. J. Chem. Inf. Model. **59**(2), 895–913 (2018)

20. Sulimov, V.B., Kutov, D.C., Taschilova, A.S., Ilin, I.S., Tyrtyshnikov, E.E., Sulimov, A.V.: Docking paradigm in drug design. Curr. Top. Med. Chem. **21**(6), 507–546 (2021)
21. Tietze, S., Apostolakis, J.: GlamDock: development and validation of a new docking tool on several thousand protein- ligand complexes. J. Chem. Inf. Model. **47**(4), 1657–1672 (2007)
22. Trott, O., Olson, A.J.: AutoDock vina: improving the speed and accuracy of docking with a new scoring function, efficient optimization, and multithreading. J. Comput. Chem. **31**(2), 455–461 (2010). https://doi.org/10.1002/jcc.21334, https://dx.doi.org/10.1002/jcc.21334
23. Wang, C., Zhang, Y.: Improving scoring-docking-screening powers of protein-ligand scoring functions using random forest. J. Comput. Chem. **38**(3), 169–177 (2017)
24. Wang, R., Fang, X., Lu, Y., Yang, C.Y., Wang, S.: The PDBbind database: methodologies and updates. J. Med. Chem. **48**(12), 4111–4119 (2005)

Search for Zinc Complexes with High Affinity in Pyrazinamidase from *Mycobacterium Tuberculosis* Resistant to Pyrazinamide

Jesus Antonio Alvarado-Huayhuaz[1,2] , Daniel Alonso Talaverano-Rojas[1] ,
Reneé Isabel Huamán Quispe[1] , Maurício Dorneles Caldeira Balboni[2] ,
Oscar Emilio Arrúa Arce[2] , Adriano Velasque Werhli[2] ,
Karina dos Santos Machado[2(✉)] , and Ana Cecilia Valderrama-Negrón[1]

[1] Laboratorio de Investigación en Biopolímeros y Metalofármacos (LIBIPMET),
Facultad de Ciencias de la Universidad Nacional de Ingeniería, Lima, Peru
jalvaradoh@uni.pe
[2] COMBI-Lab, Grupo de Biologia Computacional, Centro de Ciências
Computacionais, Universidade Federal do Rio Grande, Rio Grande, RS, Brazil
karina.machado@furg.br

Abstract. Tuberculosis is an ancient and current disease. Resistance to the prodrug pyrazinamide (PZA), one of the most important antituberculosis drug, is often associated with various mutations in the *pncA* gene that expresses the metalloenzyme pyrazinamidase (PZase). Some hard and intermediate acids, such as Co(II), Mn(II), and Zn(II), showed the ability to partially recover the susceptibility to PZA in resistant strains. In this work, we investigate the affinity that zinc complexes can achieve in the PZase protein with a low affinity for pyrazinamide. First, we select the PZase mutant with the best resistance profile to PZA using the webserver SUSPECT-PZA and a home-made script. Then we use the tmQM database, which contains 86,665 metal complexes with crystallographic structures and quantum descriptors, to search for zinc complexes with high affinity for PZase. Out of 5867 Zn complexes, 100 with lower dipole moment, higher hardness and lower HOMO energy were selected. Molecular docking studies using (a) empirical scoring functions (SF) and (b) SF based on machine learning, allowed us to find complexes such as BUXZUQ, FEQTUS or DOSQUA that have higher affinity for PZase than PZA. These Zn complexes not only exhibit higher global reactivity compared to PZA, but are also very similar to each other, and to a lesser degree their organic part is also similar to that of PZA. The compounds we have reported can serve as a basis for the design of new antituberculosis metallodrugs.

Keywords: Zn complexes · Pyrazinamidase · Molecular docking · Machine learning

N. M. Scherer and R. C. de Melo-Minardi (Eds.): BSB 2022, LNBI 13523, pp. 109–120, 2022.
https://doi.org/10.1007/978-3-031-21175-1_12

1   Introduction

Tuberculosis (TB) is a disease caused by *Mycobacterium tuberculosis* (*Mtb*). Although it is an ancient disease, it still causes a high mortality rate worldwide, currently ranking in the top 10. The number of new cases has increased in recent years due to the COVID-19 pandemic, which is an alarming scenario for the coming decades [11]. The first-line drugs for the treatment of tuberculosis are rifampicin, isoniazid, ethambutol, and pyrazinamide (PZA), and one of the reasons for the failure of TB treatment is resistance caused by the evolution of bacterial strains [29].

From these first-line drugs, Pyrazinamide (PZA) is one of the main as it can shortening TB therapy and kill persistent strains that other drugs cannot, for this reason, it is recommended by the WHO, even in the treatment of multidrug-resistant TB [32]. PZA is converted to pyrazinoic acid (POA), in the cytoplasm of *Mtb* by hydrolysis with the enzyme pyrazinamidase (PZase). POA is the active species against *Mtb* and has the ability to cross the mycobacterium membrane in a repetitive cycle of entry (by passive diffusion, like H-POA) and exit (by efflux system yet to be identified, like POA anion). In this way, it acts as a proton internalizer, which acidifies the cytoplasm, accumulates POA and alters homeostasis in permeability and transport through the membrane, leading to cell death [27]. Other pharmacological targets in *Mtb* are ribosomal protein S1 (RpsA) and the aspartate decarboxylase enzyme (PanD). PZA resistance genes have been reported in *panD*, *rpsA* and mainly in *pncA* of *Mtb* encoding PZase, however, the mechanisms of action and of resistance to PZA in Mtb are incompletely understood [34].

In wild type PZase, (a) ASP8, LYS96, and CYSs138 form the catalytic triad; (b) TRP68 and PHE13 are the substrate binding residues and; (c) ASP49, HIS51 and HIS71 are the iron binding amino acids. Various mutations in (b), (c) and neighboring amino acids are associated with resistance to PZA, due to the alterations to bind to the substrate or due to the loss of the Fe2+ cation that affects the catalytic cycle to produce POA. Because of this PZA resistance in Mtb, new drug candidates are needed today [19].

In vitro studies show the recovery of susceptibility to PZA when PZase coordinates with Co(II), Mn(II) or Zn(II) [24,26]. Du et al. [8] had obtained crystals of PZase with Zn(II) in *Pyrococcus Horikoshii*. Similarly, crystals of Zn(II):Fe(II) 1:1 were obtained in PZase from *Acinetobacter baumannii* and PZase from *Mtb* with Fe(II) [10]. According to recent studies, some metallochaperones, such as Rv2059, could be the *in vivo* deliverers of Zn(II) [27]. Since PZase activity depends on metal ions, some studies have reported new metallodrugs with potential antituberculosis activity [6,7,17,22]. At the catalytic site, the metal-oxygen distance decreases with increasing metal charge, i.e., with increasing Pearson hardness. This leads to an increase in dipolar moment, acidity, and probably production of POA in the catalytic cycle [23]. If the Zn cation can improve the susceptibility of the *pncA* mutant, it is interesting to know what binding affinity it can reach in the active site of the PZase and how its mode of interaction is.

Protein-ligand molecular docking technique is widely used to predict the binding/conformation modes of a small molecule at its target receptor binding site and to estimate the affinity of this complex. Scoring Functions (SF) are functions used to estimate the binding energy of each pose, and can be classified into four categories: physics-based, empirical, knowledge-based and machine-learning-based [16,28]. There is a particular interest in machine learning based SF because they offer improvements in predictions compared to the other classes of SF [2,28,31]; in general, these models are proposed using a large number of descriptors and trained using methods that are not necessarily linear [16].

RFL-Score is a machine-learning-based scoring function inspired by various works available in the literature [14], its training set is made up of 5273 experimental complexes and 3773 decoys obtained from public databases, it uses 160 molecular descriptors as input data calculated using open-source software (Biophyton, DSSP, BiNANA, PaDEL-Descriptor, RDKit 2D/3D, MSMS, Autodock Vina); its model is trained using the Random Forest algorithm of the Scikit-learn software, where the hyperparameters max_depth, max_features and n_estimators are optimized with GridSearchCV; finally, RF-Score is validated using CASF-2016, obtaining the following results: 0.812 R (Scoring Power), 0.696 p (Ranking Power), 86.7% (Docking Power) and 28.1% (Screening Power) [3].

Therefore, in this work we investigate the affinity that zinc complexes can achieve in the PZase-mutated protein with a low affinity for PZA. Ligands were selected from a database of transition metal complexes and the PZase-mutated protein from SUSPECT-PZA [12]. Docking simulations were performed using Schrodinger 2021-3 suite and binding affinity was assessed using Glide and RFL score. The proposed methodology and results can serve as a starting point for the search for metallodrugs based on zinc complexes against resistant strains of *Mtb*. The paper is organized as follows: Sect. 2 describes the material and methods, Sects. 3 and 4 present and discuss the results, and Sect. 5 draws conclusions and presents future work.

2 Material and Methods

Figure 1 summarizes the proposed methodology for this work.

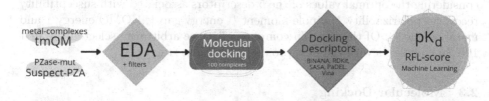

Fig. 1. Flow chart of the proposed methodology. tmQM: transition metal Quantum Mechanics Database, EDA: exploratory data analysis, Kd: dissociation constant.

2.1 Obtaining the Data

The first step is to obtain the structures of the receptor and ligands. The structure of the wild type PZase (Pzase-wt) was obtained from the Protein Data Bank (3PL1, 2.2 Å) [20] and that of the mutant PZase (PZase-mut) was obtained using the web-server SUSPECT-PZA [12].

To select Pzase-mut, we use an in-house developed script to search PZase mutant candidates in SUSPECT-PZA varying key amino acids in the binding site and possible mutations. This script compiles all the output parameters of the SUSPECT-PZA in a CSV file for each mutation and downloads the structures in PDB format of the proteins. Of the mutant proteins with better susceptibility/resistance indicators, the one with lower affinity (more positive docking score) for PZA than PZase-mut was selected. The chosen PZase-mut 3D structure (F13G) is available at github.com/combilab-furg/Jesus, as well as the Python codes used for the exploratory data analysis (EDA) and others on this step.

The set of Zn complexes was obtained from the transition metal Quantum Mechanics (tmQM) data set [4]. tmQM comprises 86,665 transition metal-organic mononuclear complexes extracted from Cambridge Structural Database (CSD), including Werner, bioinorganic and organometallic complexes with closed-shell, formal charge in the range $\{+1, 0, -1\}$e, Cartesian coordinates, optimized structures at the GFN2-xTB level [5] and eight quantum descriptors: Electronic energy, (2) Dispersion, (3) Dipole Moment, (4) Metal Charge, (5) Highest Occupied Molecular Orbital Energy, (6) Lowest Unoccupied Molecular Orbital Energy, (7) HOMO-LUMO Energy ou energy gap and (8) Polarizability.

Since PZA is not found within tmQM, its optimized 3D structure was calculated with Gaussian09 with B3LYP/6-31G(d,p), in implicit aqueous solvent. This material is available at doi.org/10.19061/iochem-bd-6-150.

2.2 Exploratory Data Analysis About Ligands

First, of the 86,665 tmQM complexes, we selected all the Zn complexes (5,867). Then, we used *Python* libraries (Pandas, Numpy, Matplotlib and Statsmodels) to visual inspection and preprocessing. This step is important to understand the descriptors distribution of this data set. Finally we selected Zn complexes considering the optimal values of the 5 descriptors associated with susceptibility recovery (polarizability ↓, dipole moment ↓, energy gap ↓, HOMO energy ↓ and metal charge ↑). Of the final Zn complexes list, we arbitrarily selected the first 100 Zn complexes with the best profile for the next stage.

2.3 Molecular Docking

The Schrodinger 2021-3 suite was used for the molecular docking simulations [1,9,21,33]. The receptors Pzase-wt and PZase-mut structures were pretreated using Protein Preparation Wizard. Crystallized waters were removed from the

Pzae-WT and a pH of 6.0 was adjusted for both structures with PropKa (according to the *in vitro* studies performed for *Mtb*). In addition, the energy of the system was minimized with the force field OPLS4. The box was built using Receptor Grid Generator, centralized in the catalytic triad, using an inner box of 10 Å and an outer box of 10 Å. Formal charges on the coordination bonds were performed manually with the *zero-order bonds to metal* module. Since the Zn complexes do not require structural optimization because they are crystallized structures, they were not treated in LigPrep. Glide in extra precision mode was used for docking the Zn complexes and the Ligand Interaction Diagram module to visualize the types of intermolecular interactions.

2.4 Docking Descriptors and pK_d RFL-Score

The final complexes (Zn complex - PZase-mut) obtained from docking simulations were converted to mol2 (for the Zn complexes and the control PZA) and PDB/FASTA (for the PZase and PZase-Mut) file formats. These files were used to generate 160 molecular descriptors (Binana, RDKit 2D and 3D, PaDEL, Delta Vina, etc.) and predict the negative logarithm of the protein-ligand dissociation constant (pKd) using RFL-score. This SF was developed in our research group and is available at github.com/combilab-furg/rfl-score_v1 [3]. Finally, Lisica was used to calculate the similarity index between the ligands [15].

3 Results

3.1 Exploratory Data Analysis for Ligands Selection

For an overview about the quantum descriptors of the total Zn complexes from tmQM, these data were normalized and plotted. The violin plot shows that the data density (y-axis) of each property is concentrated in one or at most two numerical values (Fig. 2). HOMO and LUMO mainly populate intermediate values and the high data dispersion in HL_gap facilitates the selection stage of complexes with higher global reactivity.

The correlation matrix presented in Fig. 2 shows a higher positive correlation when it tends toward red (1.00) and a higher negative correlation when it tends toward blue (−1.00). We found a strong negative correlation between polarizability and electron dispersion (−0.97). This indicates that if, in the population of Zn complexes in tmQM, we find molecules with low polarizability, it is very likely that they present high dispersion, that is, resonant effects due to diffuse electron clouds (π conjugates) [30], as in the case of PZA. We found a good positive correlation (>0.55) between LUMO energy versus the HOMO and HL_Gap descriptors. That is, Zn complexes in tmQM, with low LUMO, are more likely to present low HOMO and low HL_Gap, therefore, they are expected to present high global reactivity.

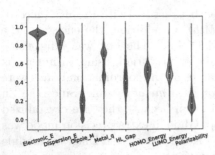 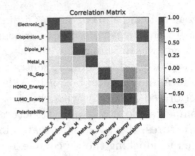

Fig. 2. Violin plots (left) and correlation matrix (right) of normalized Zn complexes descriptors.

3.2   Molecular Docking Simulations

F13G mutation in PZase (PZase-mut) was selected among more than 300 mutations generated in the active site, due to a set of susceptibility/resistance indicators returned by SUSPECT-PZA. The affinity between PZA and PZase-wt (−4.318 kcal/mol) is reduced when compared with the PZA and PZase-mut (−3.262 kcal/mol), probably caused by the disappearance of the pi-stacking interaction between PHE13 and the PZA pyrazine ring.

PZA best pose predicted by docking in the PZase-mut is located in the vicinity of the iron (Fig. 3a), in a predominantly apolar region (Fig. 3b). The hydrophobic amino acids VAL7, LEU19, ILE133, ALA134 and CYS138 mainly stabilize the ligand, although we also found two hydrogen bond receptors, in ILE133 and ASP8, from the amide group (Fig. 3c). In Fig. 3d we present the contribution in the interaction energy that goes from −7.37 in blue to 1.38 kcal/mol in red, in this way ILE133 and ASP8 stand out. Analogously in Figs. 3e, 3f, and 3g, we identify the importance of ILE133 and ASP8, again, in the other intermolecular interactions.

3.3   Scoring Values: Glide and RFL-Score

We selected the docking results according to Glide BE scores lesser than −6 kcal/mol (or the highest affinity ligand-protein), totalling nine Zn complexes. The identifier codes in CSD are: BUXZUQ, EHIBIG, QERMIK, DOSQUA, FEQTUS, AVOQUX, FIDWUL, UHETUY and FIDWOF (Table 1, columns 1 and 3). Considering only the organic part of these complexes, we found some similarity between the ligands with PZA, mainly in AVOQUX, BUXZUQ and DOSQUA (Table 1, column 4). It is also observed that most of the selected compounds have a lower E gap (Table 1, column 5), which indicates that they require less energy to promote an electronic transition between the frontier orbitals, that is, they have a higher global reactivity when comparted to PZA.

In order to compare Glide and RFL-Score we normalized the score values obtained considering PZase-mut as receptor. For RFL-Score, the minimum predicted pKd was for PZA (2.34) and maximum for the BUXZUQ (4.94) while

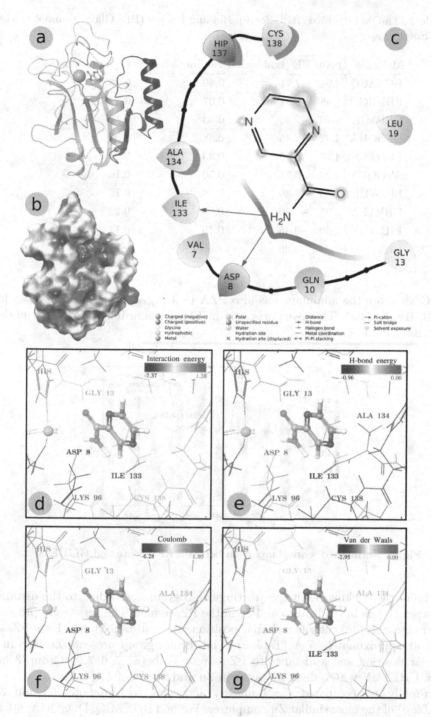

Fig. 3. PZA - PZase-mut: ligand interaction diagrams. These diagrams were generated by Schrodinger 2021-3 suite.

Table 1. TmQM hits: pKd (RFL-Score), Binding Energy (BE) Glide, Tanimoto Index and Energy gap.

Molecule	pKd	BE (kcal/mol)	Tanimoto index	E-gap (hartrees)
BUXZUQ	4.94	−7.71	0.40	0.08
EHIBIG	4.48	−6.81	0.07	0.20
QERMIK	4.40	−7.95	0.33	0.11
DOSQUA	4.32	−6.45	0.38	0.09
FEQTUS	4.27	−6.89	0.33	0.11
AVOQUX	4.23	−6.71	0.50	0.16
FIDWUL	3.53	−6.36	0.14	0.15
UHETUY	3.21	−6.40	0.00	0.22
FIDWOF	2.69	−6.46	0.15	0.14
PZA	2.34	−3.26	1.00	0.18*

for Glide score the minimum was also PZA (−3.26) and the maximum was for QERMIK (−7.95). This comparison is in Fig. 4 where it is possible to notice proximity in most cases.

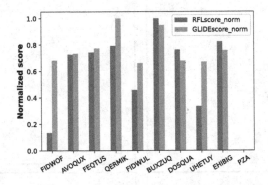

Fig. 4. Normalized score comparison between RFL-Score and GLIDE SFs.

From the docking results, we distinguish 3 groups according to the distance between the Zn in the ligands and Fe in the PZase-mut. The first group presents their zinc metallic center in a similar position, and a distance to the Fe of PZase-mut at approximately 5 Å (Table 2). The second group presents Zn also in a similar position and a distance to PZase-mut Fe between 6–7 Å. Group 3 has only UHETUY with a distance between Zn and the Fe of 4.362 Å.

Finally, we compared the similarity according to Tanimoto 2D and 3D (Table 3) of the most similar Zn-complexes. We find BUXZUQ, DOSQUA, QERMIK, with a similarity close to 1, according to the Tanimoto 2D index, and around 0.5, according to the Tanimoto 3D index.

Table 2. Zn-Fe distances, d(Zn-Fe), in molecular docking

Group 1		Group 2		Group 3	
Zn complex	Dist Zn-Fe	Zn complex	Dist Zn-Fe	Zn complex	Dist Zn-Fe
EHIBIG	5.081	AVOQUX	5.931	UHETUY	4.362
FIDWOF	5.198	BUXZUQ	6.363		
FIDWUL	5.007	DOSQUA	6.575		
		FEQTUS	6.871		
		QERMIK	6.133		

Table 3. 2D and 3D Tanimoto index between tmQM hits

2D	BUXZUQ	DOSQUA	QERMIK	3D	BUXZUQ	DOSQUA	QERMIK
BUXZUQ	1.00	0.94	0.83	BUXZUQ	1.00	0.67	0.65
DOSQUA	0.94	1.00	0.79	DOSQUA	0.67	1.00	0.48
QERMIK	0.83	0.79	1.00	QERMIK	0.65	0.48	1.00

4 Discussion

As reported by Sheen et al. [27], the involvement of some metallic cations such as Co2+, Mn2+ and Zn2+ can recover susceptibility to PZA. These are hard acids (Mn2+) and intermediates of Pearson (Co2+ and Zn2+), as well as the native cation of the metalloenzyme, Fe2+, which is also an acid intermediate. Since the metal charge is more positive, it polarizes the metal-oxygen bond and favors the deprotonation of the water molecule coordinated with the metal to form a nucleophile that catalyzes the enzymatic reaction, i.e., producing more POA [13,25]. From this perspective, we can justify our selection of compounds in tmQM by choosing the hardest compounds, i.e., with lower polarizability and higher metal charge. We also selected compounds with low HOMO energy, or relative metal-ligand stability and low energy gap, that is, require an amount of energy comparable to that used by PZA to produce an electronic transition. Given that the active site of PZase has a predominantly neutral electrostatic potential surface, and the vast majority of amino acids that stabilize PZA are hydrophobic, we consider selecting the Zn compounds with the lowest dipolar moment. Therefore, the selected molecules correspond to a filter with low polarizability, low dipole moment, low energy gap, low HOMO energy, and high metallic charge.

According to the RFL-score, higher predicted pKd values indicate better affinity between the ligand and the protein. Therefore, BUXZUQ is the Zn complex with the best affinity for PZase-mut, whereas it is QERMIK for Glide. Such differences are relative to the score developed by each program, so we do not expect equal numbers but rather comparable trends. For example, because our starting point for selecting the mutation in PZase was to obtain a lower affinity compared with the wild-type protein, it was expected that the pKd value predicted by RFL-score would give the lowest value of the group, as shown in Table 1. Of the nine compounds evaluated in Fig. 4, we found that seven had the expected trend, with FIDWOF and UHETUY being the exceptions, as these two compounds were expected to have a higher pKd value.

The Tanimoto indices of the organic part of the Zn complexes already give an idea of the features favoring a good coupling to the active site of the protein with respect to PZA. On the other hand, all Zn complexes found are between 23 and 29 atoms (with the exception of EHIBIG that has 35) and most contain water molecules in their first coordination sphere. This is reflected in the distances between Zn and Fe and in the 3D Tanimoto coefficients for the groups mentioned in the Results section (Tables 2 and 3).

The proximity or direct intermolecular interactions between the catalytic triad of PZase-mut and the Zn-complexes found are also important. However, the effect in this enzyme is not necessarily the conversion of the amine group to carboxylate. One possible mechanism could be the delivery of drugs coordinated with Zn for a possible synergistic effect that restores the susceptibility of the resistant strain to PZA and helps the drug to react with the catalytic triad. Considering that PZA is a prodrug that exerts its toxicity on *Mtb* mainly by altering intracellular pH produced by POA and inhibiting PanD protein, it is not exptected that these Zn-complexes selected in this work to follow this route. The mechanism used by these Zn complexes will also depend on their stability, the possible exchange of ligands with the water in the medium and, of course, the pH to which they are exposed.

5   Conclusions

In this work we selected the F13G mutation in the PZase enzyme of Mtb (PZase-mut), which is resistant to PZA, due to its reduced affinity for this drug. We proposed a methodology to select Zn complexes for PZase-mut using different datasets and tools: PDB, SUSPECT-PZA [12], Schrodinger 2021-3 suite [1,9,21,33], RFL-Score [3], *Python* and Lisica [15]. We reported nine tmQM Zn complexes with high affinity for this receptor. These complexes present greater global reactivity, greater hardness and similarity to PZA (considering only the organic part), and therefore, they could promote the recovery of susceptibility to PZA in resistant strains. These complexes can be used to design new metallodrugs against Tuberculosis.

As future work we propose the development of a repository specifically focused on metallodrugs, which does not currently exist [18]. From this perspective, this repository will be used to predict synergistic models (by physical mixture or by covalent conjugated drugs) that are currently available only through *in vitro* experiments [6,17].

Acknowledgments. The team thanks Dr. Ataualpa Carmo Braga, from the Institute of Chemistry of the University of Sao Paulo, for the support in the use of Gaussian09. JAAH thanks the financial support of the Management Agreement No. 237-2015-FONDECYT to the Vice-Rectorate for Research of the Universidad Nacional de Ingeniería of Peru. KSM, AVW, OEAA and MDCB thank the Conselho Nacional de Desenvolvimento Científico e Tecnológico (CNPq) [process number 439582/2018-0], Coordenação de Aperfeiçoamento de Pessoal de Nível Superior (CAPES) [Finance Code 001] and Fundação de Amparo a Pesquisa do Rio Grande do Sul (FAPERGS) [process number 22/2551-0000385-0].

References

1. Adeniyi, A.A., Ajibade, P.A.: Comparing the suitability of autodock, gold and glide for the docking and predicting the possible targets of RU (II)-based complexes as anticancer agents. Molecules **18**(4), 3760–3778 (2013)
2. Ain, Q.U., et al.: Machine-learning scoring functions to improve structure-based binding affinity prediction and virtual screening. Wiley Interdiscip. Rev. Comput. Mol. Sci. **5**(6), 405–424 (2015)
3. Arce, O.E.A.: Função de escore baseada em machine learning para docagem molecular proteína-ligante. Master's thesis (2020)
4. Balcells, D., Skjelstad, B.B.: tmQM dataset-quantum geometries and properties of 86k transition metal complexes. J. Chem. Inf. Model. **60**(12), 6135–6146 (2020)
5. Bannwarth, C., et al.: GFN2-xTB-an accurate and broadly parametrized self-consistent tight-binding quantum chemical method with multipole electrostatics and density-dependent dispersion contributions. J. Chem. Theory Comput. **15**(3), 1652–1671 (2019)
6. Chávez Llallire, N.K., et al.: Síntesis, caracterización y evaluación de la actividad biológica de compuestos de coordinación de cobalto con pirazinamida. Rev. Soc. Quim. Peru **86**(3), 315–328 (2020)
7. Coelho, T., et al.: Metal-based antimicrobial strategies against intramacrophage mycobacterium tuberculosis. Lett. Appl. Microbiol. **71**(2), 146–153 (2020)
8. Du, X., et al.: Crystal structure and mechanism of catalysis of a Pyrazinamidase from Pyrococcus horikoshii. Biochemistry **40**(47), 14166–14172 (2001)
9. Friesner, R.A., et al.: Extra precision glide: docking and scoring incorporating a model of hydrophobic enclosure for protein- ligand complexes. J. Med. Chem. **49**(21), 6177–6196 (2006)
10. Fyfe, P.K., et al.: Specificity and mechanism of Acinetobacter baumanii nicotinamidase: implications for activation of the front-line tuberculosis drug pyrazinamide. Angew. Chem. Int. Ed. **48**(48), 9176–9179 (2009)
11. Jeremiah, C., et al.: The who global tuberculosis 2021 report - not so good news and turning the tide back to end TB. Int. J. Infect. Dis. (2022)
12. Karmakar, M., et al.: Structure guided prediction of pyrazinamide resistance mutations in pncA. Sci. Rep. **10**(1), 1–10 (2020)
13. Khadem-Maaref, M., et al.: Effects of metal-ion replacement on pyrazinamidase activity: a quantum mechanical study. J. Mol. Graph. Model. **73**, 24–29 (2017)
14. Kundu, I., et al.: A machine learning approach towards the prediction of protein-ligand binding affinity based on fundamental molecular properties. RSC Adv. **8**(22), 12127–12137 (2018)
15. Lesnik, S., et al.: LiSiCA: a software for ligand-based virtual screening and its application for the discovery of butyrylcholinesterase inhibitors. J. Chem. Inf. Model. **55**(8), 1521–1528 (2015)
16. Liu, J., Wang, R.: Classification of current scoring functions. J. Chem. Inf. Model. **55**(3), 475–482 (2015)

17. Maldonado, Y.D., et al.: Evaluation of their potential as prospective agents against mycobacterium tuberculosis. J. Inorg. Biochem. **227**, 111683 (2022)

18. Medina-Franco, J.L., et al.: Bridging informatics and medicinal inorganic chemistry: toward a database of metallodrugs and metallodrug candidates. Drug Discov. **27**(5), 1420–1430 (2022)

19. Njire, M., et al.: Pyrazinamide resistance in mycobacterium tuberculosis: review and update. Adv. Med. Sci. **61**(1), 63–71 (2016)

20. Petrella, S., et al.: Crystal structure of the Pyrazinamidase of mycobacterium tuberculosis: insights into natural and acquired resistance to pyrazinamide. PLoS One **6**(1), e15785 (2011)

21. Prasad, H.N., et al.: Design, synthesis and molecular docking studies of novel piperazine metal complexes as potential antibacterial candidate against MRSA. J. Mol. Struct. **1232**, 130047 (2021)

22. Quaresma, S., Alves, P.C., Rijo, P., Duarte, M.T., André, V.: Antimicrobial activity of pyrazinamide coordination frameworks synthesized by mechanochemistry. Molecules **26**(7), 1904 (2021)

23. Rasool, N., Husssain, W., Khan, Y.D.: Revelation of enzyme activity of mutant Pyrazinamidases from mycobacterium tuberculosis upon binding with various metals using quantum mechanical approach. Comput. Biol. Chem. **83**, 107108 (2019)

24. Rasool, N., Iftikhar, S., Amir, A., Hussain, W.: Structural and quantum mechanical computations to elucidate the altered binding mechanism of metal and drug with Pyrazinamidase from mycobacterium tuberculosis due to mutagenicity. J. Mol. Graph. **80**, 126–131 (2018)

25. Salazar-Salinas, K., et al.: Metal-ion effects on the polarization of metal-bound water and infrared vibrational modes of the coordinated metal center of mycobacterium tuberculosis Pyrazinamidase via quantum mechanical calculations. J. Phys. Chem. B **118**(34), 10065–10075 (2014)

26. Sheen, P., et al.: Role of metal ions on the activity of mycobacterium tuberculosis Pyrazinamidase. Am. J. Trop. Med. Hyg. **87**(1), 153 (2012)

27. Sheen, P., et al.: Metallochaperones are needed for mycobacterium tuberculosis and Escherichia coli Nicotinamidase-Pyrazinamidase activity. J. Bacteriol. **202**(2), e00331–19 (2020)

28. Shen, C., et al.: From machine learning to deep learning: advances in scoring functions for protein-ligand docking. Wiley Interdiscip. Rev. Comput. Mol. Sci. **10**(1), e1429 (2020)

29. Singh, R., et al.: Recent updates on drug resistance in mycobacterium tuberculosis. J. Appl. Microbiol. **128**(6), 1547–1567 (2020)

30. Smith, Q.A., Ruedenberg, K., Gordon, M.S., Slipchenko, L.V.: The dispersion interaction between quantum mechanics and effective fragment potential molecules. J. Chem. Phys. **136**(24), 244107 (2012)

31. Su, M., et al.: Comparative assessment of scoring functions: the CASF-2016 update. J. Chem. Inf. Model. **59**(2), 895–913 (2018)

32. Sun, Q., et al.: The molecular basis of pyrazinamide activity on mycobacterium tuberculosis panD. Nat. Commun. **11**(1), 1–7 (2020)

33. Vijayakrishnan, P., Antony, S.A., Velmurugan, D.: Structural data of DNA binding and molecular docking studies of dihydropyrimidinone transition metal complexes. Data Br. **19**, 817–825 (2018)

34. Zhang, Y., et al.: Mechanisms of pyrazinamide action and resistance. Microbiol. Spectr. **2**(4), 2–4 (2014)

How Bioinformatics Can Aid Biodiversity Description: The Case of a Probable New Species of *Orthonychiurus* (Collembola, Hexapoda)

Maithe Gaspar Pontes Magalhaes[1], Marilia Alves Figueira Melo[1],
Gabriel Costa Queiroz[2] (iD), Aline dos Santos Moreira[1] (iD), Wim Degrave[1] (iD),
and Thiago Estevam Parente[1](✉) (iD)

[1] Laboratório de Genômica Funcional e Bioinformática, IOC, Fiocruz, Rio de Janeiro, RJ, Brazil
thiago.parente@fiocruz.br
[2] Museu Nacional do Rio de Janeiro, UFRJ, Rio de Janeiro, RJ, Brazil

Abstract. The description of all living species is an ultimate goal of biology. Species description, however, is a time-consuming effort that requires specialized taxonomists in the vast array of existing taxa. With the current rate of habitat loss, it is plausible to assume more species are becoming extinct than what we could possibly describe. High-throughput sequencing together with the appropriate bioinformatic analyses is revolutionizing the knowledge regarding microbial biodiversity, but can also assist in the unraveling of the diversity of higher Eukaryotes. Here, we describe how transcriptome sequencing, *de novo* assemble, and BLAST analyses helped to identify the taxa of an unknown and abundant species sampled on the banks of a highly contaminated river in Rio de Janeiro, Brazil. This species is currently being described as a possible new species of Collembola (Hexapoda). In total, 4.589.437 paired-end 150 bp reads passed quality control and were used to assemble a de novo transcriptome, resulting in 44.013 transcripts with N50 of 1.338 bp. Of these assembled transcripts, 4.112 had a BLAST hit, with *Folsomia candida* (Collembola) being the most frequent species. Specimens of this sampled species were sent to a taxonomist specialized in Collembola for accurate taxonomic identification. The sampled specimens are being fully described and probably belong to a new species of *Orthonychiurus* (Hexapoda, Collembola, Onychiuridae).

Keywords: High-throughput sequencing · Transcriptome · Species identification

1 Introduction

The fundamental question about how many species currently exist on Earth still has no answer [1, 2]. In Anthropocene Era, however, answering this question is more urgent than ever as the rate of species extinctions is exacerbated, causing the extinction of species that have never and now will never be known to humankind [3, 4]. Forever ignoring these species and their genetic resources halts not only our understanding of how life evolved and diversified in this planet, but also our ability to discover new and useful molecules

N. M. Scherer and R. C. de Melo-Minardi (Eds.): BSB 2022, LNBI 13523, pp. 121–127, 2022.
https://doi.org/10.1007/978-3-031-21175-1_13

and genomic or metabolic pathways that could benefit humankind from improvements in the treatment of medical conditions to the discovery of new biotechnological processes.

Traditionally, the discovery of new species relies on heavy sampling strategies that are frequently taxon-oriented and carried out by taxon-specific experts. After sampling, the shelf life between discovery and description has been estimated to be 21 years [5]. Thus, it is urgent the need for a more efficient strategy to discover and describe new species. The current availability of high throughput sequencing and the array of bioinformatic tools offers an unprecedented opportunity to optimize the efforts to discover and expedite the description of new species. Here, we describe a successful example of a sample-to-sequence strategy that generated abundant transcriptomic data from an unidentified specimen, allowing the fast and accurate submission of the sample to the appropriate taxon-specific taxonomist, and the ongoing description of a new species.

2   Material and Methods

2.1   Sampling and RNA Extraction

The sequenced species was sampled on the banks of Cunha Canal (22°52'52.9"S 43°14'28.2"W) Rio de Janeiro, Brazil, as a by-catch during fish sampling for another project (ICMBio permit number 75704-1; SISGEN registration number AAB444D). The abundance of "tiny little worms" among a bunch of mud from this highly and multi-contaminated canal caught our attention. Due to the large volume of untreated sewage from the surrounding settlements discharged into the Cunha Canal, we supposed right after sampling that these "tiny little worms" could be a kind of human intestinal parasite; so, we decided to preserve few individuals in RNALater in an attempt to identify the specimens species.

After examining few individuals under a stereoscope, our initial supposition was ruled out, but we had still almost no clue about what species was that. We then decided to extract total RNA from a whole individual for transcriptome sequencing. Total RNA was isolated using the phenol-chloroform method, with TRIzol reagent. The isolated RNA was quantified using NanoDrop. The quality and integrity of RNA were analyzed using TapeStation.

2.2   Transcriptome Sequencing and Analysis

The sequencing libraries were prepared using the ILMN Strnd mRNA Prep kit with individual indexes (IDT ILMN RNA UDI A) and 150 bp paired-end sequencing was performed on the HiSeq2500 at the Fiocruz NGS Facility RTP01J. The sequencing files were preprocessed in BaseSpace (Illumina), where fastq files were generated and adapters trimmed.

The fastq files were transferred to the Fiocruz Bioinformatics Core Facility RPT04A. The reads were filtered by quality using the Trimmomatic v.0.39 with the following parameters: LEADING:28 TRAILING:28 SLIDINGWINDOW:4:28 MINLEN:36. Reads quality was analyzed using FastQC v0.11.9. The *de novo* transcriptome was assembled using the default parameters of Trinity V2.11.0. The coding sequences were extracted from the Trinity assembled transcripts using the default parameters of Trans-decoder. The sequenced reads were mapped using the default parameters of Bowtie2 against the complete set of CDS, and only those CDS with 50 or more reads mapped were kept for further analysis. This selected set of CDS were subjected to BLASTn searches against the complete NCBI nr database. Only the BLASTn top hit was retrieved. Using scripts developed in-house, the species and the gene names were retrieved from the entry description. A list of unique species names was used to count the number of entries for each species. Likewise, a list of unique gene names was used to count the number of entries for each gene. The transcripts annotated as belonging to the most frequent species were retrieved from the transcriptome and subjected to another round of mapping using the default parameter of Bowtie2.

2.3 Taxonomical Confirmation

The other sampled individuals were sent to a taxonomist expert in the biological group suggested by the BLASTn results. Stereomicroscopical analysis before and after depigmentation has been performed and a complete description of this putative new species is ongoing.

3 Results

The data described here is deposited in U.S.A. NCBI databases with the following accession number: PRJNA860273 (BioProject), SAMN29837419 (BioSample), and SRR20324169 (SRA).

3.1 Taxa Assignment to an Unidentified Specimen Through Transcriptome Analysis

The whole-organism transcriptome sequencing of a single individual of an unidentified species generated more than 7 million reads, of which almost 5 million passed quality control and were used to *de novo* assemble more than 44 thousand transcripts, with an average length of 838 bp (Table 1). In this newly assemble transcriptome, more than 27 thousand protein coding sequences (CDS) were identified (Table 1). These CDS were all subjected to BLASTn against the NCBI nr database, resulting in more than 3 thousand hits (Table 1). The most frequent species among these BLAST top hits was *Folsomia candida* Willem 1902 (Hexapoda, Collembola, Isotomidae) with 174 entries, which had an average nucleotide identity of 76% and an average alignment length of 686 nucleotides (Table 1). Although the number of entries assigned to the most frequent species represents only 5% of the total BLAST hits, 70% of the reads used to assemble the transcriptome mapped in those 174 transcripts assigned to *F. candida* (Table 1). The CDS not assigned to *F. candida* were assigned to hundreds of other species, including a few other Collembolan species, and together mapped 26% of the reads used in the assemblage. It is believed most of these reads originated from RNA content from food sources still available in the digestive tract of the sampled specimen. The low percentage identity between the assembled transcripts and their respective *F. candida* hits suggested the sampled specimen is a different species, while *F. candida* being the most frequent species, and the high percentage of reads mapped against BLAST hits assigned to *F. candida* indicates the sampled specimen is a Collembola. Therefore, the sampled specimens were sent to an expert Collembola taxonomist for species identification.

Table 1. Summary of *Orthonychiurus* sp. nov. (Collembola, Onychiuridae) transcriptome sequencing and annotation.

Total sequencing reads	7,282,265
Reads after QC	4,589,437
Trinity	
Genes	36,197
Transcripts	44,013
N50	1,338
Average length (bp)	838
CDS	27,554
BLAST hits	3,158
Average identity	78%
Average length (bp)	550
Top species BLAST hits	174
Average identity	76%
Average length (bp)	686
Mapping	
CDS (all)	96%
CDS assigned to top species	70%

3.2 Species Investigation

The sampled specimen was observed under an appropriate stereomicroscope before and after depigmentation (Fig. 1) and confirmed to be a Collembola. The designated ge-nus, however, is not *Folsomia candida*, a worldwide distributed isotomid species frequently used as a model organism in ecotoxicological tests. Instead the genus was conformed as *Orthonychiurus*, a cosmopolitan genus of Onychiuridae. Due to its usage in laboratory studies, *F. candida* is by far the collembolan species most represented in the existing nucleotide databases. Therefore, the assignment of *F. candida* as the most probable species comes with no surprise after BLAST analysis of any unknown Collembola species. According to the Checklist of the Collembola of the World [6] there are no known species of *Orthonychiurus* registered around the geographical area where these specimens were sampled. Accordingly, the sampled specimens do not fully match the diagnostic criteria for the already described species of *Orthonychiurus*. The sampled specimen appears to be a new species of *Orthonychiurus* and is currently being investigated for a complete species description.

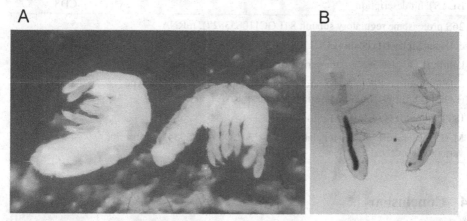

Fig. 1. Probable new species of *Orthonychiurus* (Collembola). Photomicrography under light stereomicroscope of representative individuals of the sequenced. A. Specimens preserved in ethanol with original coloration; B. Unpigmented specimens in lactic acid (gut contents are evident due to transparency of the cuticle).

3.3 Evidence of Gene-Specific Expansions in *Orthonychiurus sp. nov.*

The 174 CDS assembled in the *Orthonychiurus sp. nov.* transcriptome shared 117 gene descriptions, indicating that different transcripts were annotated as the same gene (Table 2). The maximum number of CDS assigned to the same gene description was 5. In the case of Kalirin, the 5 assembled CDS showed different lengths (1,500; 1,596; 1,944; 3,303; and 3,651 bp). Kalirin is known to have differently spliced isoforms with varying lengths [7, 8]. Likewise, other 5 assembled CDS with different lengths (1,872; 3,510; 4,398; 4,812; and 4,938 bp) were annotated as Collagen alpha-1, which is also known to have differently spliced isoforms with varying lengths [9]. Differently, the 5 assembled CDS annotated as 26S proteasome regulatory subunit 8 had the same length, 1,203 bp. Table 2 shows the annotations assigned to four or more of the assembled CDS.

Table 2. Descriptions of the BLAST hits assigned to more than four CDS assembled from the *Orthonychiurus sp. nov.* (Collembola) transcriptome

BLAST hit description	CDS
26S proteasome regulatory subunit 8 (LOC110853827), mRNA	5
collagen alpha-1(I) chain (LOC110856657), mRNA	5
kalirin (LOC110858376), transcript variant X3, mRNA	5
calcium-transporting ATPase sarcoplasmic/endoplasmic reticulum type (LOC110851249), transcript variant X4, misc_RNA	4
IST1 homolog (LOC110851880), mRNA	4
MAP/microtubule affinity-regulating kinase 3 (LOC110859347), transcript variant X2, mRNA	4

4 Conclusions

Whole-body transcriptome sequencing and analysis from an initially unidentified organism expedited the classification of this specimen into a high hierarchy biological taxon, allowing the submission of this specimen to an experienced taxonomist in the group to a final species identification or, in this case, to the description of a new species. We argue that transcriptome sequencing is an efficient tool to complement the current efforts on genome and DNA barcode sequencing for the discovery and the description of new species. Transcriptome sequencing generates as much data as genome sequencing, at a fraction of the cost, while still almost as simple to perform and analyze as DNA barcode. Database biases toward model species and species from the northern hemisphere is a current major drawback preventing a more precise species identification using assembled transcripts. However, the only way to tackle this issue and improve the representation of non-model species from developing areas of the world is to sample and sequence species native to these areas. The current abundance of undescribed species worldwide, but especially in biodiversity hotspots which are frequently located in developing countries, ensures the success of a sample-to-sequence strategy like the one adopted here.

References

1. Larsen, B.B., Miller, E.C., Rhodes, M.K., Wiens, J.J.: The quarterly review of biology inordinate fondness multiplied and redistributed: the number of species on earth and the new pie of life (2017)
2. Mora, C., Tittensor, D.P., Adl, S., Simpson, A.G.B., Worm, B.: How many species are there on earth and in the ocean? PLoS Biol. **9** (2011). https://doi.org/10.1371/journal.pbio.1001127
3. Dirzo, R., Raven, P.H.: Global state of biodiversity and loss. Annu. Rev. Environ. Resour. **28**, 137–167 (2003). https://doi.org/10.1146/annurev.energy.28.050302.105532
4. Storch, D., Šímová, I., Smyčka, J., Bohdalková, E., Toszogyova, A., Okie, J.G.: Biodiversity dynamics in the Anthropocene: how human activities change equilibria of species richness. Ecography **2022** (2022). https://doi.org/10.1111/ecog.05778

5. Fontaine, B., Perrard, A., Bouchet, P.: 21 years of shelf life between discovery and description of new species (2012). https://doi.org/10.1016/j.cub.2012.10.029
6. Bellinger, P.F., Christiansen, H.E., Janssens, F.: Checklist of the Collembola of the World (2009)
7. Miller, M.B., Yan, Y., Wu, Y., Hao, B., Mains, R.E., Eipper, B.A.: Alternate promoter usage generates two subpopulations of the neuronal RhoGEF Kalirin-7. J. Neurochem. **140**, 889–902 (2017). https://doi.org/10.1111/jnc.13749
8. Mcpherson, C.E., Eipper, B.A., Mains, R.E.: Genomic organization and differential expression of Kalirin isoforms (2002)
9. Peltonen, S., Rehn, M., Pihlajaniemi, T.: Alternative Splicing of Mouse a 1 (XIII) Collagen RNAs Results in at Least 17 Different Transcripts, Predicting cd (XIII) Collagen Chains with Length Varying Between 651 and 710 Amino Acid Residues. Mary Ann Liebert, Inc. (1997)

Phylogeny Trees as a Tool to Compare Inference Algorithms of Orthologs

Rafael Oliveira[1], Saul de Castro Leite[1,2] (iD),
and Fernanda Nascimento Almeida[1,3(✉)] (iD)

[1] BHIG - Bionformatics and Health Informatics Group - CECS/UFABC,
São Bernardo do Campo, Brazil
[2] Center for Mathematics, Computing and Cognition, UFABC, Santo André, Brazil
[3] Post-Graduation Program of Biomedical Engineering - PPG-EBM, CECS,
UFABC, São Bernardo do Campo, Brazil
fernanda.almeida@ufabc.edu.br

Abstract. Orthologous genes are defined as genes arising from speciation events, being highly conserved in form and function. Several algorithms seek to identify them, but a simple methodology is not available to determine the quality of their results. This work proposed using
the definition of orthologs and the analysis of phylogenetic trees to
develop a methodology to compare these algorithms. Thirty proteomes
of prokaryotes were obtained, focusing on Leifsonia and Clavibacter genera. The orthogroups were inferred using five graph-based algorithms
(OMA, Orthofinder, PorthoMCL, ProteinOrtho, and Sonic Paranoid).
Frequencies of each homologous group were obtained from the resulting
raw data. The sequences were aligned by MUSCLE software. After that,
the sequences were trimmed by the trimAl software and concatenated
into supermatrices. The percentage of information for each supermatrix was calculated. The phylogenetic trees were built applying three
tree reconstruction methods: Maximum Likelihood, Bayesian inference,
and Neighbors-joining. The reference trees were made by 16S ribosomal
RNA sequences. Furthermore, gene trees from orthogroups with *taxa =
30* were inferred by the Maximum Likelihood methodology. The trees
were compared to the reference tree by topology and Robinson-Foulds
distances. Despite the differences in the quantity of the orthogroups
obtained from each algorithm, no significant differences were observed
between the constructed trees. However, previous work with other distinct species verified that this methodology may be viable. It is concluded that the proposed methodology is valid, although not to all species
groups. Due to the input data dependencies, this methodology is recommended to be performed for each new data set.

Keywords: Algorithms · Orthologous genes · Phylogenetic trees

1 Introduction

Homologous characteristics are the result of ancestry shared between different
taxa. To understand how homologous genes are related, the term has been refined

N. M. Scherer and R. C. de Melo-Minardi (Eds.): BSB 2022, LNBI 13523, pp. 128–139, 2022.
https://doi.org/10.1007/978-3-031-21175-1_14

to molecular systematics studies. The terms orthologs and paralogs mean, respectively, genes that have a common ancestry and were separated by speciation events and genes that originated from duplication events. Genes from horizontal transfer are called xenologs. Gene similarity is a measure commonly used in methodologies that seek to predict the function and structure of genes. It is useful, for example, to compare and characterize newly sequenced genes, and that can be done using orthologous genes. By definition, orthologs tend to retain equivalent functions in different species, as they are the most similar in terms of sequence, structure, domain, and function [15].

As they represent speciation nodes, orthologous genes are powerful tools for constructing phylogenies. The same does not apply to paralogous genes, which not correctly represent the evolutionary history of the group [1]. After a duplication event, both genes would likely share the same function, but selective pressures tend to negatively select one or both genes or diverge them in function [17]. In literature, it is possible to find several examples of phylogenies based on orthologs, highlighting the importance of this type of data [14]. Other applications, such in Healthcare, Biotechnology, and Pharmacology, benefit from their conservative aspect. They facilitate the discovery of new molecular targets, as functions of an ortholog can be extrapolated to their homologs, including genes from different species. TDR Targets, for example, is a biological database that, among its categories, includes orthologous genes for use in molecular and biochemical studies [32].

Many studies rely on the distinction between orthologs and paralogs, raising the need for tools to interpret and analyze genomic data. Several computational algorithms are developed annually to identify orthogroups, i.e.., all genes that descend from the last common ancestor of a clade of species and originated from speciation events [22]. Methodologies are commonly divided into two categories: Graph-based and tree-based. There is no consensus on which one is the best method. Graph-based methods are based on pairwise genes comparisons across species, and the least divergent sequences are called orthologs [3]. These algorithms typically have two steps, the construction in which pairs of orthologs are inferred by similarity score and are connected to edges, and the clustering where orthogroups are constructed based on the graph structure. Gene similarity score can be done by bidirectional best hit (BBH) [24], which considers the common similarity of alignments, or by reciprocal shortest distance (RSD) [33], which measures their phylogenetic distance. Tree-based algorithms use gene trees with internal nodes labeled as speciation or duplication events, to determine which genes are orthologs and paralogs [3]. These input trees are obtained by reconciling gene and species trees by parsimony strategies. Tree-based methods result in more information with high statistical power by working simultaneously with all sequences. However, they need more processing power and depend on reconciled trees.

Due to its importance for comparative biology, several methods have been developed to infer orthogroups. Widely used graph-based methods include OMA [9], Orthofinder [12], PorthoMCL [30], ProteinOrtho [20], and Sonic Paranoid [7], each with an approach to overcoming challenges in orthologs discovery such as duplication events, horizontal gene transfers, insertions and deletions. Therefore, different behaviors are expected according to the input data, making the choice of algorithm very important. OrthoMCL, for example, has been reported as problematic when working with plant genomes due to problems identifying duplication events [14].

The development of comparative methodologies became necessary to assess the performance of these algorithms. Since 2010 the Quest for Orthologs (QfO) consortium has maintained an online benchmarking service, Orthology Benchmarking (http://orthology.benchmarkservice.org), to compare orthology methods. The service provides reference proteomes, tests based on phylogeny and gene function, and data from public submissions [2]. The benchmarking system is carried out with annually manually cured reference orthogroups [2]. Its results represent how well the tested methodology manages to retrieve these reference groups with precision and recall rates. Although annual updates of the QfO corrects inconsistencies, it is crucial to have a reliable independent methodology where the user can define which data will be tested. Emms and Kelly [13], for example, verified that 39% of the original reference orthogroups [31] were incorrect.

Phylogenetic trees are essential for comparative studies in biology, including genomics. As orthologs represent speciation nodes, it is valid to state that when comparing phylogenetic trees constructed by orthogroups obtained from different algorithms, it is possible to infer their quality. Previous work with this methodology applied to genomes of fungi, mollusks, and arachnids does not indicate a consensus regarding this approach [4,28]. When it comes to prokaryotic genomes, the only robust data regarding the comparison of results from inference algorithms of orthologs are found in the Orthologs - Orthology Benchmarking [2]. In this work, we seek to validate the use of only phylogenetic analysis to compare the performance of different inference algorithms to orthologous genes. For this, phylogenetic trees built with the orthogroups resulting from each algorithm were compared.

2 Methodology

To reduce the computational cost, in this work were used only prokaryote proteomes. Thirty species proteomes were selected from the NCBI genome database. We choose to analyze proteomes of Leifsonia and Clavibacter genera because of our group focus on these species. Other species were chosen without specific criteria, including bacteria from the Microbacteriaceae family and Gram-negative bacteria as an external group. From all select species, 16S ribosomal RNA (16S rRNA) sequences were also obtained from the SILVA database for construction of the reference trees (ReT) (https://www.arb-silva.de/) [16].

Orthogroups were inferred by five graph-based algorithms listed in Table 1. This category was chosen for its low computational consumption and for not depending on reconciled trees. Unless otherwise stated, configurable parameters from each algorithm have been set as default to compare their standard runs. We performed the algorithms on a computer with the following configurations: CPU Intel®Core™ i5 8250U (1.60 GHz), 8 GB RAM (DDR4), 1 TB hard disk, and Linux Ubuntu v.19.

Table 1. Inference algorithms for orthogroups used in this work.

Algorithm	Version	Observation	Reference
OMA	2.3.1	It was inferred OMA Groups/HOGS	Dessimoz et al. 2005 [9]
Orthofinder	2.3.3	BLAST + MSA was used as default	Emms and Kelly 2019 [12]
PorthoMCL	1.0	None	Tabari and Su 2017 [30]
ProteinOrtho	6.0	None	Lechmer et al. 2011 [20]
Sonic Paranoid	1.2.6	Due to memory limitations, only half of the available threads were used	Cosentino and Iwasaki 2018 [7]

Frequencies of each type of homolog groups were obtained from the raw output data. For descriptive purposes, co-orthologs were considered paralog groups. Single copy orthogroups with more than 300 amino acids (AAs) and present in more than 50% of the species ($taxa \geq 15$) were selected for phylogenetic analysis. These sequences were aligned by MUSCLE software [11] in default parameters. Poorly aligned regions were eliminated by trimAl software with -automated1 flag [6]. These steps were repeated for the 16S rRNA sequences. The processed sequences were concatenated into supermatrices. The equation (1) was used to calculate the percentage of information (Info%) for each tree supermatrix (with exception of the ReT supermatrix), defined as the ratio between the number of sites with amino acids (AAS) and the total number of sites (TS), including gaps, of the tree supermatrix.

$$Info(\%) = \frac{AAS}{TS}.100 \qquad (1)$$

The phylogenetic trees, including the ReT, were built applying different tree reconstruction methods to increase the method robustness. The RAxML v.8.2.12 [29] was used for the Maximum Likelihood of multiple aligned sequences method, with GTRGAMMA and PROTCATLG substitution models for ReT and orthogroups trees, respectively, both with a convergence of 1000 bootstraps (#autoMRE). Bayesian inference was performed by MrBayes v.3.2 [27] using the default parameters (generations = 20000 sample frequency = 500, print frequency = 500, diagnostic frequency = 5000 and default run-length = 1000000) and default substitution models. For the Neighbor-Joining method, MEGA X [19] was used with default settings, 1000 bootstraps, and distance calculation by the Poisson correction method. Furthermore, for each single-copy orthogroup with $taxa = 30$, the sequences were aligned and trimmed by the respective software previously mentioned, and used to construct gene trees by the Maximum Likelihood method with the previous settings.

The Robinson-Foulds (RF) distance [26] between the ReT and the orthogroups/gene trees was calculated by RaxML v.8.2.12. Means and standard deviations were calculated for the gene trees RF distances. Significance for differences between means was verified by the ANOVA test with the R aov() function. For the gene trees, the mean RF distances were compared with the data available in Quest for Orthologs - Orthology Benchmarking, Species Tree Discordance Benchmark (2018 public data) for Bacteria, Eukaryota and Fungi.

Qualitative comparisons between the nodes of the phylogenetic trees were performed using the software FigTree v.1.4.4 (http://tree.bio.ed.ac.uk/software/figtree/).

3   Results and Discussion

To compare the algorithms presented in Table 1, first we analyzed the quantitative parameters of the raw output data. From the total generated data (Fig. 1a), OMA Groups produced 18,352 homolog groups followed by ProteinOrtho (15,895), PorthoMCL (11,027), Sonic Paranoid (11,526), OMA HOGs (10,974), and Orthofinder (9,223). Phylogenetic analyses are usually based on gene families where no duplication events occurred, like in orthogroups. Other types of homolog groups, such as paralogs and co-orthologs, tend to make analysis difficult because the evolutionary history of the gene is not always congruent with the evolutionary history of the species [17].

Only orthogroups with one-to-one genes (single copy) were used for the final analyses. From the total orthogroups (Fig. 1a), the previous proportion was maintained with OMA Groups inferring 18,352 groups, followed by ProteinOrtho with 14,044 groups and the remaining algorithms below the mean value. When we observed the total paralog groups distribution (Fig. 1a), the proportion was inverted concerning the total orthogroups. The algorithms that inferred the smallest amount of orthogroups are now the most influential paralog groups, 3,425 paralog groups to Orthofinder, 3,268 to PorthoMCL, and 3,009 to Sonic Paranoid. In a practical sense, this indicates less helpful information at the end of the process. For example, approximately 37% of the groups inferred by Orthofinder are not used in the final analysis. The OMA Groups did not produce any paralog group, so all the data generated in its pipeline was used in the phylogenetic construction.

An essential aspect of phylogenetic trees inference is the quality of the input data. In traditional phylogeny, matrices containing homologous characteristics shared by large groups of species are preferable to matrices with characteristics present only in small groups of species, because it decreases the probability of missing data that can lead to polytomies. A reduced number of polytomies, which may represent missing or ambiguous data, is desirable to a better results interpretation. Gaps are defined as sites with no information (missing data) after sequence alignment. They can be created by sequences size differences derived from base insertion or deletion. They can also occur due to gene absence in some species in the analyzed population. So, the selection of orthogroups for a phylogenetic analysis must pay attention to the number of species represented in each one to reduce gene absence. This work used orthogroups with at least 50% of information present ($taxa \geq 15$) to reduce the number of gaps and utilize as much information as possible without compromising the phylogenetic analysis.

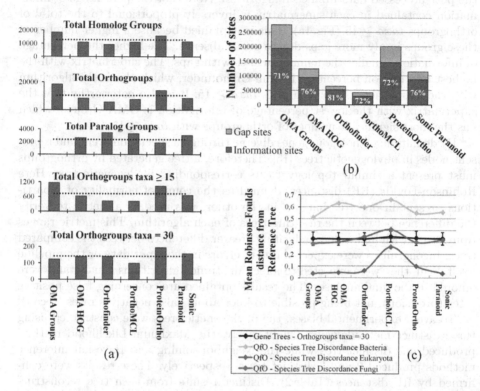

Fig. 1. (a) Quantitative description of the results from each algorithm (dotted line represents the mean value). (b) Characterization of each supermatrix. (c) Curves of the mean Robinson-Foulds distances of Gene Trees. Orthogroups $taxa = 30$ (black), QfO - Species Tree Discordance Bacteria (yellow), Eukaryota (blue) and Fungi (red). Bars - Standard deviation. For QfO Bacteria, Eukaryota, and Fungi standard deviation are below 0.02. (Color figure online)

Comparing the total orthogroups with the total orthogroups $taxa \geq 15$ (Fig. 1a) a similar proportion were observed. This was expected because more orthogroups imply more groups with high species representativeness. However, restricting the observation to orthogroups $taxa = 30$, the pattern changes (Fig. 1a). Despite having the highest total orthogroups, the OMA Groups are the second smallest in groups with $taxa = 30$, only behind PorthoMCL. A valuable classification for orthologs predictor algorithms is their precision and recall rates [1]. Higher precision implies a lower number of false positives, while higher recall decreases the rate of false negatives [5]. The OMA Groups have high precision and low recall, which explains the smaller number of groups with more species representation [4]. The other algorithms are near the Pareto frontier, the definition given to a set of optimal solutions with their respective trade-offs in multiobjective scenarios, in this case, egalitarian precision and recall rates [2].

Figure 1b presents the characteristics of the supermatrices produced with the post-processed data from each algorithm. Note that the percentage of information contained in each supermatrix is inversely proportional to the total of orthogroups $taxa \geq 15$ (Fig. 1a). This is explained because a higher number of these groups imply more gaps due to genes absence. The higher the percentage of information, smaller the number of sites with gaps. The supermatrix with the highest information percentage is from Orthofinder, which also is the algorithm with the least amount of orthogroups $taxa \geq 15$. It is also noteworthy that the supermatrix with the lowest percentage of information is OMA Groups, which has the most significant number of orthogroups with $taxa \geq 15$.

By definition, orthologous genes diverge through a speciation event and represent nodes in phylogenetic trees [15]. Therefore, a tree generated by orthogroups must present a similar topology to its corresponding phylogenetic tree. Here Robinson-Foulds (RF) distances, defined as the count of normality of bipartitions present in one tree but not in the other, was used as a metric to verify the difference between the resulting trees of each algorithm. This metric ranges from 0 to 1, with high values indicating greater differences between the compared trees. Incongruities were expected due to errors in the input data or the pipeline itself. Still, they were expected to affect all studied algorithms equally and were ignored in the final analyses. The results obtained are concerning false positive rates (precision), not being possible to infer about false negatives rates (recall).

To avoid experimental biases, the phylogenetic trees were constructed using three distinct methodologies. Qualitatively, the Maximum Likelihood method produced one distinct tree while the Neighbor-joining and Bayesian inference methods produced 2 and 3 distinct trees, respectively. These results were confirmed by RF distances (Table 2). Distinct results from each tree reconstruction method were expected, because different substitution models can produce discordant results. It has also been described that the accuracy of these tree reconstruction methods varies according to input data [34], with the number of ambiguous characters, such as gaps, being an important factor to differentiate their results.

Nevertheless, the tree reconstruction method does not appear to be significant to impact this study results. Individually the RF distances from each tree reconstruction method remained constant, except for the Orthofinder and PorthoMCL algorithms for the Bayesian inference and Neighbor-joining methods (Table 2). Therefore, for the other algorithms, it is possible to affirm that the lack of distinct topologies between the trees implies no significant difference regarding the quality of their generated data.

For Bayesian inference results, the RF distances from Orthofinder and PorthoMCL differ 3.7% and 7.4% from the other algorithms, respectively, representing approximately 1 and 2 distinct internal nodes (Fig. 2). The trees from both algorithms are identical in the Neighbor-joining method, with their RF

Table 2. Robinson-foulds distances from reference tree and the orthogroups trees by each algorithm/methodologies.

	Maximum Likelihood	Bayesian Inference	Neighbor Joining	Average value
OMA Groups	0.259	0.296	0.296	0.284 ± 0.021
OMA HOG	0.259	0.296	0.296	0.284 ± 0.021
OrthoFinder	0.259	0.259	0.259	0.259 ± 0.00
PorthoMCL	0.259	0.222	0.259	0.247 ± 0.021
ProteinOrtho	0.259	0.296	0.296	0.284 ± 0.021
Sonic Paranoid	0.259	0.296	0.296	0.284 ± 0.021

distances varying 3.7% compared to the other algorithms (Table 2). Orthofinder and PorthoMCL have the smallest supermatrices, but this factor alone does not explain the more significant similarity of their trees with ReT. Interestingly, the PorthoMCL tree construct by Bayesian inference was the only one that correctly inferred the relationship of Clavibacter subspecies [8]. These results are likely the effect of a combination of factors, such as the supermatrices characteristics, the quality of the data, and the tree reconstruction method used. However, the low RF distances from the Orthofinder and PorthoMCL trees do not allow us to affirm any aspects regarding the quality of their generated data since this occurred in only two tree reconstruction methods, and this bias cannot be disregarded.

Each supermatrix had different characteristics in terms of percentage of information and gap sites. Despite being essential factors in phylogenetic analysis, the amount of data and gap sites do not always have a deterministic role in the quality of the final result. The influence of the gaps can vary according to the data set worked on [18]. The quality of the input data prevails over its quantity since the increase of non-informative sequences contributes little to the phylogenetic analysis [25]. Many high-quality phylogenetic analyses are performed using only one target gene with few nucleotides, as with 16S rRNA sequences. However, despite the differences previously verified for the precision and recall rates of the used algorithms [1,2], which affects the quality of the orthogroups, such metrics played no significant role in our final results. Deutekom and Snel [10] demonstrate that even distinct orthogroups can result in the same phylogeny.

For each orthogroups $taxa = 30$, a gene tree was constructed by the Maximum Likelihood method. RF distances were averaged for each set of gene trees. No significant differences were observed between the means (p ¿ 0.05). The mean values of RF distances of gene trees were compared with those obtained in QfO - Orthology Benchmarking for Bacteria, Eukaryota, and Fungi 2018 data sets [2] (Fig. 1c). The chosen benchmark was Species Tree Discordance Benchmark, whose methodology is similar to the one performed in this study. For the RF distances of the PorthoMCL, data from OrthoMCL [21] were used.

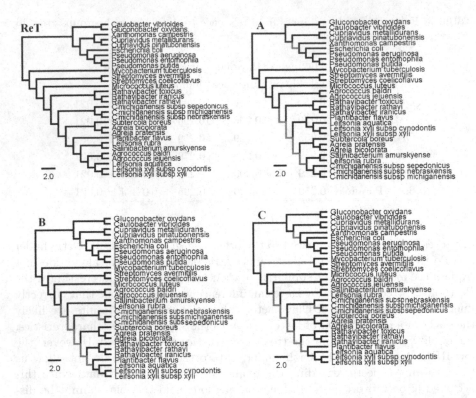

Fig. 2. Phylogenetic trees built with Bayesian inference method. ReT - Reference Tree. A - PorthoMCL. B - Orthofinder. C - Other algorithms. Scale - replacements for every hundred sites. In red - diverging branches. (Color figure online)

As noted, the QfO values differ from those observed in this study in both magnitude and mean variation. Comparing the results, it can be seen that the curve of the gene trees has a linear behavior, which does not occur for the others. The QfO curves, except for Bacteria, can be approximated by a linear expression removing the PorthoMCL component. PorthoMCL achieved the highest RF distance compared to other algorithms, implying great incongruity with its ReT. Based on these results, it is possible to state that for this dataset (gene trees and QfO curves), PorthoMCL was the algorithm with the worst performance.

The results obtained for both orthogroups and gene trees suggest that the proposed methodology is not sensitive enough to detect differences in evolutionarily close groups. Several of the selected species belong to the same family (Microbacteriaceae) or are closely related (Leifsonia and Clavibacter genera) [23]. Phylogeny reconstructions at the level of genus, order, etc., as performed in this study, may not be affected by the presence of paralog groups [14]. More diverse species, as in QfO, can facilitate the identification of differences that algorithms of orthologs can cause in the topology of phylogenetic trees, as the complexity of these data makes it more difficult to infer phylogenetic relationships. QfO utilizes a set of reference proteomes that are manually cured or derived from trusted sources [1,31]. Species diversification is a feature present in these reference databanks.

The sensitivity of the proposed methodology may be directly related to the input data, either in quality or complexity. Several works with similar approaches reported favorable and unfavorable results for this method. Shen et al. [28] compared the performance of 3 inference algorithms of orthologs using 332 budding yeast genomes, resulting in phylogenies with approximately 10% differences (32 nodes). Altenhoff et al. [4] also verified incongruous topologies when reconstructing the Lophotrochozoa phylogeny using five distinct inference algorithms. On the other hand, Kallal et al. [18] used two inference algorithms to reconstruct the phylogeny of Araneidae family, and no significant differences were identified in the topologies. Therefore, it is impossible to determine which algorithms of orthologs is the best for every application. So, it is recommended that this methodology be performed for each new data set, preferably combined with other types of analysis, such as those based on conservation of gene function, for additional information.

4 Conclusion

Despite the quantitative differences in the results, there were no significant differences in the phylogenetic analyses. Therefore, for the group of bacteria used in this work, the inference algorithm of orthologs was not significant for the phylogenetic reconstruction. However, other works obtained different phylogenies when constructed by orthogroups of different algorithms [4,28]. The group of species analyzed, in addition to other factors, can be a decisive factor in the accuracy of the algorithm of orthologs. Evolutionarily closer groups of species, such as those in this work, may provide orthogroups with greater reliability compared to groups with evolutionarily distant species.

References

1. Altenhoff, A.M., et al.: Standardized benchmarking in the quest for orthologs. Nature Methods **13**(5), 425–430 (2016). https://doi.org/10.1038/nmeth.3830
2. Altenhoff, A.M., et al.: The quest for orthologs benchmark service and consensus calls in 2020. Nucleic Acids Res. **48**(1), 538–545 (2020). https://doi.org/10.1093/nar/gkaa308
3. Altenhoff, A.M., Glover, N.M., Dessimoz, C.: Inferring orthology and paralogy. In: Anisimova, M. (ed.) Evolutionary Genomics. MMB, vol. 1910, pp. 149–175. Springer, New York (2019). https://doi.org/10.1007/978-1-4939-9074-0_5
4. Altenhoff, A.M., Levy, J., et al.: OMA standalone: orthology inference among public and custom genomes and transcriptomes. Genome Res. **29**(7), 1152–1163 (2019). https://doi.org/10.1101/gr.243212.118
5. Buckland, M., Gey, F.: The relationship between recall and precision. J. Am. Soc. Inf. Sci. **45**(1), 12–19 (1994). https://doi.org/10.1002/(SICI)1097-4571(199401)45:1⟨12::AID-ASI2⟩3.0.CO;2-L

6. Capella-Gutiérrez, S., Silla-Martínez, J.M., et al.: trimAl: a tool for automated alignment trimming in large-scale phylogenetic analyses. Bioinformatics **25**(15), 1972–1973 (2009). https://doi.org/10.1093/bioinformatics/btp348

7. Cosentino, S., Iwasaki, W.: SonicParanoid: fast, accurate and easy orthology inference. Bioinformatics **35**(1), 149–151 (2019). https://doi.org/10.1093/bioinformatics/bty631

8. Michael, J., Davis, A., Gillaspie, G., et al.: Clavibacter: a new genus containing some phytopathogenic coryneform bacteria, including clavibacter xyli subsp. xyli sp. nov., subsp. nov. and clavibacter xyli subsp. cynodontis subsp. nov., pathogens that cause ratoon stunting disease of sugarcane and bermudagrass stunting disease. Int. J. Syst. Evol. Microbiol. **34**(2), 107–117 (1984). https://doi.org/10.1099/00207713-34-2-107

9. Dessimoz, C., et al.: OMA, a comprehensive, automated project for the identification of orthologs from complete genome data: introduction and first achievements. In: McLysaght, A., Huson, D.H. (eds.) RCG 2005. LNCS, vol. 3678, pp. 61–72. Springer, Heidelberg (2005). https://doi.org/10.1007/11554714_6

10. Deutekom, E.S., Snel, B., et al.: Benchmarking orthology methods using phylogenetic patterns defined at the base of eukaryotes. Briefings Bioinf. **22**(3), bbaa206 (2021). https://doi.org/10.1093/bib/bbaa206

11. Edgar, R.C.: MUSCLE: multiple sequence alignment with high accuracy and high throughput. Nucleic Acids Res. **32**(5), 1792–1797 (2004). https://doi.org/10.1093/nar/gkh340

12. Emms, D.M., Kelly, S.: OrthoFinder: phylogenetic orthology inference for comparative genomics. Genome Biol. **20**(1), 238 (2019). https://doi.org/10.1186/s13059-019-1832-y

13. Emms, D.M., Kelly, S.: Benchmarking orthogroup inference accuracy: revisiting orthobench. Genome Biol. Evol. **12**(12), 2258–2266 (2020). https://doi.org/10.1093/gbe/evaa211

14. Fernández, R., Gabaldón, T., Dessimoz, C., et al.: Orthology: Definitions, Inference, and Impact on Species Phylogeny Inference (2019). https://arxiv.org/abs/1903.04530

15. Gabaldón, T., Koonin, E.V.: Functional and evolutionary implications of gene orthology. Nat. Rev. Genetics **14**(5), 360–366 (2013). https://doi.org/10.1038/nrg3456

16. Oliver Glöckner, F., Yilmaz, P., et al.: 25 years of serving the community with ribosomal RNA gene reference databases and tools. J. Biotechnol. **261**, 169–176 (2017). https://doi.org/10.1016/j.jbiotec.2017.06.1198

17. Hellmuth, M., Wieseke, N.: From sequence data including orthologs, paralogs, and xenologs to gene and species trees. In: Pontarotti, P. (ed.) Evolutionary Biology, pp. 373–392. Springer, Cham (2016). https://doi.org/10.1007/978-3-319-41324-2_21

18. Kallal, R.J., Fernández, R., et al.: A phylotranscriptomic backbone of the orbweaving spider family araneidae (Arachnida, Araneae) supported by multiple methodological approaches. Mol. Phylogenet. Evol. **126**, 129–140 (2018). https://doi.org/10.1016/j.ympev.2018.04.007

19. Kumar, S., Stecher, G., et al.: MEGA X: molecular evolutionary genetics analysis across computing platforms. Mol. Biol. Evol. **35**(6), 1547–1549 (2018). https://doi.org/10.1093/molbev/msy096

20. Lechner, M., Findeiß, S., Steiner, L., et al.: Proteinortho: detection of (Co-)orthologs in large-scale analysis. BMC Bioinf. **12**(1), 124 (2011). https://doi.org/10.1186/1471-2105-12-124

21. Li, L., Stoeckert, C.J., et al.: OrthoMCL: identification of ortholog groups for eukaryotic genomes. Genome Res. **13**(9), 2178–2189 (2003). https://doi.org/10.1101/gr.1224503
22. Nichio, B.T., Marchaukoski, J.N., Raittz, R.T.: New tools in orthology analysis: a brief review of promising perspectives. Frontiers Genet. **8**, 165 (2017). https://doi.org/10.3389/fgene.2017.00165
23. Nordstedt, N.P., Roman-Reyna, V., et al.: Comparative genomic understanding of gram-positive plant growth-promoting leifsonia. Phytobiomes J. **5**(3), 263–274 (2021). https://doi.org/10.1094/PBIOMES-12-20-0092-SC
24. Overbeek, R., Fonstein, M., D'souza, M., et al.: The use of gene clusters to infer functional coupling. In: Proceedings of the National Academy of Sciences, vol. 96, no. 6, pp. 2896–2901 (1999). https://doi.org/10.1073/pnas.96.6.2896
25. Philippe, H., Brinkmann, H., et al.: Resolving difficult phylogenetic questions: why more sequences are not enough. PLOS Biol. **9**(3), e1000602 (2011). https://doi.org/10.1371/journal.pbio.1000602
26. Robinson, D.F., Foulds, L.R.: Comparison of phylogenetic trees. Math. Biosci. **53**(1), 131–147 (1981). https://doi.org/10.1016/0025-5564(81)90043-2
27. Ronquist, F., Teslenko, M., et al.: MrBayes 3.2: efficient bayesian phylogenetic inference and model choice across a large model space. Syst. Biol. **61**(3), 539–542 (2012). https://doi.org/10.1093/sysbio/sys029
28. Shen, X.X., Opulente, D.A., et al.: Tempo and mode of genome evolution in the budding yeast subphylum. Cell **175**(6), 1533-1545.e20 (2018). https://doi.org/10.1016/j.cell.2018.10.023
29. Stamatakis, A.: RAxML version 8: a tool for phylogenetic analysis and post-analysis of large phylogenies. Bioinformatics **30**(9), 1312–1313 (2014). https://doi.org/10.1093/bioinformatics/btu033
30. Tabari, E., Zhengchang, S.: PorthoMCL: parallel orthology prediction using MCL for the realm of massive genome availability. BigData Analytics **2**, 4 (2017). https://doi.org/10.1186/s41044-016-0019-8
31. Trachana, K., Larsson, S.P., et al.: Orthology prediction methods: a quality assessment using curated protein families. Bioessays **33**(10), 769–780 (2011). https://doi.org/10.1002/bies.201100062
32. Landaburu, L., Berenstein, A., Videla, S., et al.: TDR Targets 6: driving drug discovery for human pathogens through intensive chemogenomic data integration. Nucleic Acids Res. **48**(D1), D992–D1005 (2020). https://doi.org/10.1093/nar/gkz999
33. Wall, D.P., Fraser, H.B., Hirsh, A.E.: Detecting putative orthologs. Bioinformatics **19**(13), 1710–1711 (2003). https://doi.org/10.1093/bioinformatics/btg213
34. Yoshida, R., Nei, M.: Efficiencies of the NJp, maximum likelihood, and bayesian methods of phylogenetic construction for compositional and noncompositional genes. Mol. Biol. Evol. **33**(6), 1618–1624 (2016). https://doi.org/10.1093/molbev/msw042

Water Pollution Shifts the Soil and Fish Gut Microbiomes Increasing the Circulation of Antibiotic Resistance Genes in the Environment

Maithe Gaspar Pontes Magalhaes, Marilia Alves Figueira Melo,
Aline dos Santos Moreira⑩, Wim Degrave⑩, and Thiago Estevam Parente⁽✉⁾ ⑩

Laboratório de Genômica Funcional e Bioinformática, IOC, Fiocruz, Rio de Janeiro, RJ, Brasil
thiago.parente@fiocruz.br

Abstract. The impact of anthropogenic activities on urban rivers is responsible for changing the diversity and composition of aquatic species and microorganisms. In this study, soil and fish (*Poecilia reticulata*) faeces were sampled from two sites: the Cunha Canal, a heavily impacted urban river in Rio de Janeiro, and a reference clean site, to investigate how the Cunha Canal pollution impacts the microbiota composition and the circulation of antibiotic resistance genes (ARGs). In total, sequencing reads summed 3 million for the faeces and 9 million for the soil metagenomes. 67% of the soil and 83% of the fish faeces microbiota at both sites were classified as Bacteria. The soil microbiota of the reference site was enriched with bacteria of the genus *Bradyrhizobium* and *Streptomyces* known to fix nitrogen and to metabolize organic material, while the soil microbiota of Cunha Canal was enriched with *Acidovorax* and *Dechloromonas* known to degradate pollutants as iron and benzene. The five ARGs detected in the faeces microbiome from Cunha Canal are different from the five found in the faeces from the reference site. 22 ARGs were found in the soil sample of Cunha Canal, while no one was detected in the soil sample of the reference site. These results show that water pollution changes the microbiota diversity and increases the ARGs circulating in the environment.

Keywords: Metagenome · Urban rives · Water pollution

1 Introduction

Urban rivers are impacted by anthropogenic activities, especially by the discharge of domestic sewage and industrial contaminants, which can affect the diversity of microorganisms and promote the spread of antibiotic resistance genes (ARGs) [1]. The Cunha Canal, located in the city of Rio de Janeiro, RJ, Brazil, is one example of an urban river that suffered severe degradation by human activities. This river is surrounded by industries, slums, and one of the main avenues of Rio de Janeiro.

The associated microbiota is crucial for host development and health and in the sustainability of natural water ecosystems [2]. The continuous assessment of the impacts of

N. M. Scherer and R. C. de Melo-Minardi (Eds.): BSB 2022, LNBI 13523, pp. 140–146, 2022.
https://doi.org/10.1007/978-3-031-21175-1_15

water pollution on microbial communities of the urban rivers is relevant to public health policymakers. The aims of this work are to analyze how the pollution of Cunha Canal impacts the diversity and composition of microbial communities and the circulation of antibiotic resistance genes in the environment. To achieve these goals, the metagenomes of *Poecilia reticulata* feaces, an invasive fish species in Brazil popularly known as guppy, and the soil of Cunha Canal were sequenced and analyzed in comparison to fish feaces and soil from a reference site.

2 Material and Methods

2.1 Sampling and DNA Extraction

Male *Poecilia reticulata* and soil were sampled at Cunha Canal (22°52'51.3"S 43°14'26.1"W), a heavily polluted site on the side of the Fiocruz in Rio de Janeiro, Brazil, and at Perdido River, a clean site inside the Tijuca National Park (22°55'48.7"S 43°16'07.8"W). Fish sampling and handling were authorized by the ICMBio (licenses 75704-1 and 75868-1) and by the institutional ethical committee on the use of animals (CEUA-IOC, L-027/2019), and the access to genetic material was registered at SISGEN (AAB444D). Three fish from each site and soil were sampled. The feaces were collected and clustered according to the sample site to the total DNA extraction using the QIAamp PowerFecal Pro DNA (Qiagen). The quantification and quality analyses of soil and feaces total DNA were performed at TapeStation (Agilent).

2.2 Metagenome Sequencing and Analysis

DNA libraries were prepared using the Illumina DNA-Prep kit and individualized using IDT ILLUMINA DNA/RNA index kit. The 150 bp paired end reads of metagenomes were sequenced at Illumina MiSeq using the MiSeq® Reagent Kit v2 (300 Cycles). The reads quality control, *P. reticulata* DNA removal, and the contigs assembling were performed using the ATLAS pipeline v2.3.0 [3], with metaSpades assembler. Kraken2 v2.1.2 [4] was used for the taxonomic classification, and Resfinder 4.1 [5] for antibiotic resistance genes searching.

3 Results

The data described here are deposited in U.S.A. NCBI BioProject (PRJNA864791) with the SRA accession numbers SRR20731319, SRR20731320, SRR20731321, SRR20731322.

3.1 Metagenomes Statistics

The *Poecilia reticulata* feaces metagenome samples had 3,849,152 sequenced reads for the reference site, and 5,841,422 reads for the Cunha Canal. The soil metagenome samples had 10,173,434 sequenced reads for the reference site and 9,686,094 for the Cunha Canal. After quality control, about 94% of reads survived and were used to metagenomes assemble. A total of 9,476 and 3,237 contigs were assembled for the feaces metagenome of the reference and of the Cunha Canal sites, with the N50 of 2,319 and 1,280. Around 84% of these contigs were classified as belonging to Bacteria and 15% were unclassified. The soil metagenomes had 13,331 and 15,255 assembled contigs for the reference and the Cunha Canal sites, with the N50 of 5,747 and 3,815. Around 70% of these contigs were classified as Bacteria, and 30% were unclassified.

3.2 Taxonomic Classification

A total of 26 and 35 bacterial genus were identified in the feaces microbiome sampled at the reference and at the Cunha Canal sites. *Aeromonas veronii*, a human pathogen, was the unique species to be identified in both sites. Most of the bacteria species detected in the feaces of *P. reticulata* from the Cunha Canal are related to the degradation of xenobiotics and other environmental pollutants. Examples of bacteria species identified in the feaces of fish from the Cunha Canal capable to degrade pollutants are species of the genus *Gordonia* (Fig. 1). In contrast, most of the bacteria species identified in the microbiome feaces of *P. reticulata* sampled at the reference site are pathogenic, for example, *Citrobacter* and *Aeromonas* (Fig. 1).

The soil metagenomes of the reference and Cunha Canal sites had 100 and 99 bacteria genus identified. It's known that soil has an important role in pollutants degradation and facilitates the nutrient transformation [6]. The taxonomic classification of microbiota of the Cunha Canal soil follows the pattern found in the *P. reticulata* faeces of Cunha Canal

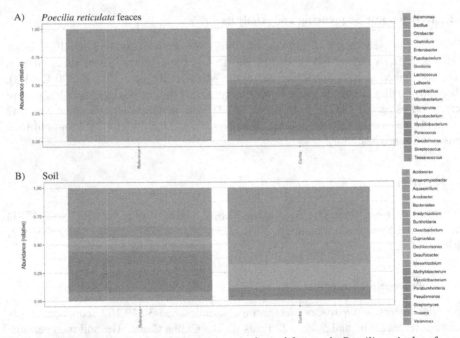

Fig. 1. Relative abundance of the 20 most common bacterial genus in *Poecilia reticulata feaces* (A) and soil (B) samples from both sites.

and are related to the degradation of aromatic compounds and other environmental contaminants, as well as with bioremediation, for example, the species of the genus *Acidovorax* and *Dechloromonas* (Fig. 1). While the ones identified at the soil of reference site includes bacterias that have significant roles in metabolize organic material and natural carbon and nitrogen cycles in freshwater, as the ones of the genus *Bradyrhizobium* and *Streptomyces* (Fig. 1).

3.3 Antibiotic Resistance Genes

The untreated wastewater discharged in urban rivers contains residual antibiotics, disinfectants, and metals that contribute to selection pressure for antibiotic resistance [7], so it is expected a higher concentration of ARGs in these impacted rivers, than in pristine sites [8]. It was identified five antibiotic resistance genes (ARGs) in the feaces metagenome of the guppy sampled in Perdido River (Table 1), five different ones in the feaces metagenome sampled in Cunha Canal (Table 1), and 22 ARGs in the soil metagenome of the Cunha Canal (Table 1), but no one in the soil metagenome of the reference site. The ARGs detected in the feaces metagenome of the *P. reticulata* sampled at Cunha Canal were also identified in the soil metagenome of the same site. These results corroborate the hypothesis that the pollution discharged in urban rivers increases the antibiotic resistance genes circulating in these environments.

Table 1. Antibiotic resistance genes identified in the sequenced metagenomes.

ARGs	Identity	Alignment/Gene length	Coverage	Accession no	Phenotype
Reference *P. reticulata* feaces					
fosA	93.9	426/426	100	AEXB01000013	Fosfomycin resistance
fosA7	93.23	325/423	76.83	LAPJ01000014	Fosfomycin resistance
cphA1	95.34	537/765	70.19	AY261379	Beta-lactam resistance
imiH	95.34	537/765	70.19	AJ548797	Beta-lactam resistance
ampH	98.24	795/795	100	HQ586946	Beta-lactam resistance
Cunha Canal *P. reticulata* feaces					
qnrS2	99.39	657/657	100	DQ485530	Quinolone resistance
mph(E)	99.45	724/885	81.81	DQ839391	Macrolide resistance

(continued)

Table 1. (*continued*)

ARGs	Identity	Alignment/Gene length	Coverage	Accession no	Phenotype
qacE	100	282/333	84.68	X68232	Disinfectant resistance
sul1	99.75	804/840	95.71	U12338	Sulphonamide resistance
aph(6)-Id	98.75	559/837	66.79	AF024602	Aminoglycoside resistance
Cunha Canal soil					
mef(A)	98.98	1179/1218	96.80	HG423652	Macrolide resistance
mef(C)	99.92	1224/1224	100	AB571865	Missing from Notes file
mph(E)	100	885/885	100	DQ839391	Macrolide resistance
mph(G)	100	885/885	100	AB571865	Missing from Notes file
msr(E)	100	1476/1476	100	FR751518	Macrolide, Lincosamide and Streptogramin B resistance
aac(6')-Ib3	99.82	555/555	100	X60321	Missing from Notes file
aadA5	99.38	648/789	82.13	AF137361	Aminoglycoside resistance
aph(6)-Id	100	837/837	100	M28829	Aminoglycoside resistance
aadA1	98.99	792/792	99.62	FJ591054	Aminoglycoside resistance
aph(3")-Ib	100	803/804	99.87	AF024602	Aminoglycoside resistance
catQ	99.24	660/660	100	M55620	Phenicol resistance
cmlA1	97.94	826/1260	65.55	AB212941	Phenicol resistance
qnrS2	100	657/657	100	DQ485530	Quinolone resistance

(*continued*)

Table 1. (*continued*)

ARGs	Identity	Alignment/Gene length	Coverage	Accession no	Phenotype
aac(6')-Ib-cr	99.42	519/519	100	EF636461	Fluoroquinolone and aminoglycoside resistance
qacE	100	282/333	84.68	X68232	Disinfectant resistance
tet(C)	99.92	1191/1191	100	AF055345	Tetracycline resistance
sul1	100	840/840	100	U12338	Sulphonamide resistance
blaVEB-1	98.63	732/900	81.33	HM370393	Beta-lactam resistance
blaOXA-9	99.15	825/825	100	KQ089875	Beta-lactam resistance
blaGES-5	100	864/864	100	DQ236171	Beta-lactam resistance
blaOXA-36	99.78	449/739	60.76	AF300985	Beta-lactam resistance
blaOXA-10	100	801/801	100	J03427	Beta-lactam resistance

4   Conclusions

The results presented here corroborate the hypothesis that the Cunha Canal pollution modulates the microbiota community present in the soil, in the fish gut, and increases the circulation in the environment of antibiotic resistance genes. It was shown that most microbial communities present in Cunha Canal are related to contaminants degradation and bioremediation, unlike what happened in the pristine river. A complete evaluation regarding the impacts of the Cunha Canal pollution on other vertebrate species, including humans, living on the banks of this urban river is necessary and will require additional efforts.

References

1. Jia, J., Gomes-Silva, G., Plath, M., Pereira, B.B., UeiraVieira, C., Wang, Z.: Shifts in bacterial communities and antibiotic resistance genes in surface water and gut microbiota of guppies (Poecilia reticulata) in the upper Rio Uberabinha, Brazil. Ecotoxicol Environ Saf [Internet] **211**, 111955 (2021). https://doi.org/10.1016/j.ecoenv.2021.111955

2. Paruch, L., Paruch, A.M., Eiken, H.G., Sørheim, R.: Faecal pollution affects abundance and diversity of aquatic microbial community in anthropo-zoogenically influenced lotic ecosystems. Sci Rep. **9**(1), 1–13 (2019)
3. Kieser, S., Brown, J., Zdobnov, E.M., Trajkovski, M., McCue, L.A.: ATLAS: a Snakemake workflow for assembly, annotation, and genomic binning of metagenome sequence data. BMC Bioinform. **21**(1), 257 (2020)
4. Wood, D.E., Lu, J., Langmead, B.: Improved metagenomic analysis with Kraken 2. Genome Biol. **20**(1), 1–13 (2019)
5. Bortolaia, V., et al.: ResFinder 4.0 for predictions of phenotypes from genotypes Valeria. J. Antimicrob Chemother (2020)
6. Ding, C., He, J.: Effect of antibiotics in the environment on microbial populations. Appl. Microbiol. Biotechnol. **87**(3), 925–941 (2010)
7. Karkman, A., Do, T.T., Walsh, F., Virta, M.P.J.: Antibiotic-resistance genes in waste water. Trends Microbiol. [Internet]. **26**(3), 220–228 (2018). https://doi.org/10.1016/j.tim.2017.09.005
8. Zhang, X.-X., Zhang, T., Fang, H.H.P.: Antibiotic resistance genes in water environment. Appl. Microbiol. Biotechnol. **82**(3), 397–414 (2008). https://doi.org/10.1007/s00253-008-1829-z

A 1.375-Approximation Algorithm for Sorting by Transpositions with Faster Running Time

Alexsandro Oliveira Alexandrino[1]([✉])[iD], Klairton Lima Brito[1][iD],
Andre Rodrigues Oliveira[1][iD], Ulisses Dias[2][iD], and Zanoni Dias[1][iD]

[1] Institute of Computing, University of Campinas (Unicamp), Campinas, Brazil
{alexsandro,klairton,andrero,zanoni}@ic.unicamp.br
[2] School of Technology, University of Campinas (Unicamp), Limeira, Brazil
ulisses@ft.unicamp.br

Abstract. Sorting Permutations by Transpositions is a famous problem in the Computational Biology field. This problem is NP-Hard, and the best approximation algorithm, proposed by Elias and Hartman in 2006, has an approximation factor of 1.375. Since then, several researchers have proposed modifications to this algorithm to reduce the time complexity. More recently, researchers showed that the algorithm proposed by Elias and Hartman might need one more operation above the approximation ratio and presented a new 1.375-approximation algorithm using an algebraic approach that corrected this issue. This algorithm runs in $O(n^6)$ time. In this paper, we present an efficient way to fix Elias and Hartman algorithm that runs in $O(n^5)$. By comparing the three approximation algorithms with all permutations of size $n \leq 12$, we also show that our algorithm finds the exact distance in more instances than the previous two algorithms.

Keywords: Genome rearrangements · Transpositions · Time complexity analysis

1 Introduction

Genome rearrangements are genetic mutations that affect one or more segments of a genome, such events are widely used in comparative genomics to estimate the evolutionary distance between organisms, which is the smallest number of rearrangement events capable of transforming one genome into another. Reversal and transposition events are among the most studied rearrangement events in the literature. The reversal inverts a segment of the genome, changing the position and the orientation of genes in the affected segment. The transposition exchanges the position of two adjacent segments of the genome.

There are different ways to represent a genome, and in the genome rearrangement field the representation by permutations is the most accepted. Each gene

N. M. Scherer and R. C. de Melo-Minardi (Eds.): BSB 2022, LNBI 13523, pp. 147–157, 2022.
https://doi.org/10.1007/978-3-031-21175-1_16

is mapped into an element of the permutation, and it is assumed that each gene is unique.

The Sorting Permutations by Transpositions problem (**SbT**) was introduced by Bafna and Pevzner [1], and the best-known approximation algorithm, presented in 2006, has an approximation factor of 1.375 [7]. In 2012, Bulteau *et al.* showed that this problem is NP-hard [2]. Even before the complexity proof, several approximation algorithms were presented for the problem [1,10], and most of them were based on a structure called cycle graph. In addition to the algorithms, heuristics have been developed to improve the practical performance [4,5]. Heuristics tend to improve the practical results, but as a drawback may end up significantly increasing the running time of the algorithms. The look-ahead is an example of such heuristics [6], which works similarly to a breadth-first search with a depth limitation parameter. The running time of the look-ahead with a minimum level of search is $\mathcal{O}(n^6)$, but it can increase depending on the value adopted for the depth limitation parameter.

The 1.375 algorithm of Elias and Hartman [7] rely on a process that transform the given permutation into a *simple permutation* and runs in quadratic time. After the transformation, the next step of this algorithm is to apply what it is called a $(2,2)$-sequence on the permutation to guarantee the approximation factor. However, a recent study [9] pointed out that this search must be done before the simplification process, since the input permutation π may have this sequence but not the simple permutation generated from π by the Elias and Hartman algorithm. The same researchers presented a new algorithm that guarantees the 1.375 approximation factor using an algebraic approach, and it makes an exhaustive search for a $(2,2)$-sequence that takes $\mathcal{O}(n^6)$ time.

In this paper we present a more efficient way to check if there is a configuration in the cycle graph that allows the use of a transposition called 2-transposition, which is necessary for the $(2,2)$-sequence. Such transposition is fundamental to obtain an algorithm for the **SbT** problem that guarantees an approximation factor of 1.375. Through this improvement, we present a $\mathcal{O}(n^5)$ algorithm for the **SbT** problem.

This work is organized as follows. Section 2 introduces the cycle graph structure and presents important definitions. Section 3 shows an efficient way to check if there is a 2-transposition in the cycle graph. Section 4 presents an improved approximation algorithm for the **SbT** problem, while Sect. 5 shows the practical results. Lastly, Sect. 6 concludes the paper.

2 Sorting Distance and the Cycle Graph

In genome rearrangement problems, we model a genome as a permutation $\pi = (\pi_1 \ \pi_2 \ \ldots \ \pi_n)$, such that each element represents a gene.

A transposition $\tau(i,j,k)$, with $1 \leq i < j < k \leq n+1$, is an operation that exchange the position of the segments $\pi_i, \ldots, \pi_{j-1}$ and $\pi_j, \ldots, \pi_{k-1}$. We represent the application of τ in the permutation π as $\pi \cdot \tau = (\pi_1 \ \ldots \ \pi_{i-1} \ \underline{\pi_j \ \ldots \ \pi_{k-1}} \ \underline{\pi_i \ \ldots \ \pi_{j-1}} \ \pi_k \ \ldots \ \pi_n)$.

The sorting distance $d(\pi)$ is equal to the minimum number of transpositions necessary to turn π into the identity permutation $\iota^n = (1\ 2\ \ldots\ n)$.

Given a permutation π, we extend this permutation by adding the elements $\pi_0 = 0$ and $\pi_{n+1} = n + 1$. The *cycle graph* of a permutation π is the undirected graph $G(\pi) = (V, E_b \cup E_g)$, where $V = \{+\pi_0, -\pi_1, +\pi_1, -\pi_2, +\pi_2, \ldots, -\pi_n, +\pi_n, -\pi_{n+1}\}$, $E_b = \{(-\pi_i, +\pi_{i-1}) \mid 1 \le i \le n + 1\}$, and $E_g = \{(-i, (i-1)) \mid 1 \le i \le n + 1\}$.

We call E_b as the set of *black edges*, which are edges that connect vertices using the position of their respective elements in π, and E_g as the set of *gray edges*, which connect vertices using the position of their respective elements in the identity permutation ι^n.

A black edge $(-\pi_i, +\pi_{i-1})$, for $1 \le i \le n+1$, has label i. When drawing the cycle graph $G(\pi)$, we follow a convention such that the vertices are placed in an horizontal line, following their positions in π. The black edges are drawn as horizontal lines such that the black edge with label 1 is the leftmost black edge and the one with label $n + 1$ is the rightmost black edge. Also, gray edges are drawn as arcs.

Since each vertex in $G(\pi)$ is incident to one black edge and one gray edge, the graph has a unique decomposition into *alternating cycles*, that is, a cycle in which any two consecutive edges are of distinct type (black edge and gray edge).

We say that an m-cycle is a cycle with m black edges and m gray edges. We say that an m-cycle is even if m is even; otherwise, we say it is odd. We represent a cycle by the list of its black edges labels: $C = (b_1, b_2, \ldots, b_m)$, where this list is constructed by traversing the cycle starting from the rightmost vertex using the black edge incident to it, that is, b_1 is the label of the black edge $(-\pi_i, +\pi_{i-1})$ that is the rightmost black edge of C and it is traversed from right $(-\pi_i)$ to left $(+\pi_{i-1})$.

The number of cycles and the number of odd cycles in the graph $G(\pi)$ is denoted by $c(\pi)$ and $c_{odd}(\pi)$, respectively. For a transposition τ, we use $\Delta c(\pi, \tau)$ to denote the change in the number of cycles caused by applying τ to π, that is, $\Delta c(\pi, \tau) = c(\pi \cdot \tau) - c(\pi)$. Similarly, we have $\Delta c_{odd}(\pi, \tau) = c_{odd}(\pi \cdot \tau) - c_{odd}(\pi)$. Figure 1 shows an example of the cycle graph built from the permutation $\pi = (5\ 4\ 1\ 6\ 3\ 2)$.

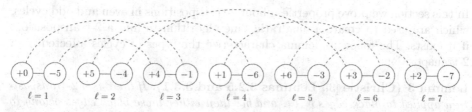

Fig. 1. Cycle graph created from the permutation $\pi = (5\ 4\ 1\ 6\ 3\ 2)$. Solid and dashed lines represent the black and gray edges, respectively. Also, notice that the black edges take a horizontal position while the gray edges form an arc over the vertices. The value ℓ, below each black edge, indicates the label assigned. (Color figure online)

In Fig. 1, the graph $G(\pi)$ has three cycles $(c(\pi) = 3)$, two cycles are even and one is odd $(c_{odd}(\pi) = 1)$. The cycles in $G(\pi)$ are: $C_1 = (3,1)$, $C_2 = (6,2,4)$, and $C_3 = (7,5)$. Note that C_1 and C_3 are even cycles while C_2 is an odd cycle.

Lemma 1 (Bafna and Pevzner [1], Lemma 2.3). *For any permutation π and transposition τ, we have that $\Delta c_{odd}(\pi,\tau) = \{-2,0,2\}$.*

Lemma 2 (Bafna and Pevzner [1], Theorem 2.4). *For any permutation π, we have that $d(\pi) \geq \frac{n+1-c_{odd}(\pi)}{2}$.*

We say that τ is a m-transposition if $\Delta c_{odd}(\pi,\tau) = m$. For instance, a 2-transposition is a transposition that increases the number of odd cycles by 2. A $(2,2)$-sequence is a sequence of 2-transpositions that can be applied consecutive to π, that is, τ,τ' is a $(2,2)$-sequence if $\Delta c_{odd}(\pi,\tau) = \Delta c_{odd}(\pi \cdot \tau, \tau') = 2$.

We classify cycles as oriented and non-oriented. A cycle $C = (b_1, b_2, \ldots, b_m)$ is *non-oriented* if $b_1 > b_2 > \ldots > b_m$. Otherwise, C is said to be *oriented*. In Fig. 1, the cycle $C_2 = (6,2,4)$ is oriented while the cycles $C_1 = (3,1)$ and $C_3 = (7,5)$ are non-oriented. Note that an oriented cycle must have at least three black edges since, in the cycle representation, we have that the label of the rightmost black edge is the first to be traversed.

Three black edges b_i, b_j, and b_k, with $i < j < k$, from the same cycle $C = (b_1, \ldots, b_m)$ form an oriented triple if one of these conditions hold: (i) $b_i > b_k > b_j$; (ii) $b_j > b_i > b_k$, or (iii) $b_k > b_j > b_i$. Bafna and Pevzner [1] showed that every oriented cycle has an oriented triple. Furthermore, they showed that a transposition τ has $\Delta c(\pi,\tau) = 2$ if, and only if, it acts on an oriented triple.

We say that an oriented triple is *valid* if the transposition acting on it is also a 2-transposition. That is, a transposition acting on a valid oriented triple increases both the number of cycles and the number of odd cycles by 2.

A *simple permutation* is a permutation such that its cycle graph has only cycles of size at most 3. Due to this characteristic, the algorithm proposed by Elias and Hartman [7] is quadratic, although it only works in permutations that were already simple permutations when given as input.

3   Finding 2-Transpositions in Quadratic Time

In this section we prove properties about 2-transpositions in even and odd cycles, which are used to create a quadratic time algorithm to find a 2-transposition, if it exists. The following lemma characterize the types of cycles affected by a 2-transposition.

Lemma 3 (Christie [3], Lemmas 3.2.5 and 3.3.1). *If there is a 2-transposition applied to black edges b_i, b_j, and b_k, then either these black edges belong to two even cycles or these three black edges belong to the same oriented cycle.*

When there are even cycles on the graph, Christie [3] showed how to find a 2-transposition in linear time. In the following Lemmas, we describe more properties of 2-transpositions affecting oriented cycles.

Lemma 4. *If there is an oriented cycle C that is even, then there is a valid oriented triple b_i, b_j, and b_k of $C = (b_1, b_2, \ldots, b_m)$, with $i < j < k$, such that $k = j + 1$.*

Proof. Bafna and Pevzner [1, Lemma 2.3] showed that every oriented cycle C has an oriented triple b_i, b_j, and b_k, with $i < j < k$, such that $k = j + 1$. A transposition applied to these black edges transform C into three cycles such that one of them is a unitary cycle. Since C is an even cycle, we know that one of the other cycles must be odd, which results in two odd cycles being added by this transposition.

Lemma 5. *If there is a valid oriented triple b_i, b_j, and b_k of an odd cycle $C = (b_1, b_2, \ldots, b_m)$ such that $i < j < k$ and $b_i > b_k > b_j$, then there is a valid oriented triple $b_{i'}$, $b_{j'}$, and $b_{k'}$ of C, with $i' < j' < k'$, such that $i' \in \{1, 2\}$ or $k' = j' + 1$.*

Proof. Note that a 2-transposition affecting only one cycle also increases the number of cycles in the cycle graph by two. This transposition creates three cycles D, D' and D'', such that D has the gray edges from the path that goes from b_i to b_j, D' has the gray edges from the path that goes from b_j to b_k, and D'' has the gray edges from the path that goes from b_k to b_i. Therefore, the size of these cycles are $|D| = j - i$, $|D'| = k - j$, $|D''| = |C| + i - k$. For a 2-transposition affecting an odd cycle, we have that $|D|, |D'|$, and $|D''|$ are odd. We divide this proof in the following cases.

When $i \in \{1, 2\}$ or $k = j + 1$ we can set $i' = i$, $j' = j$, and $k' = k$ as the valid oriented triple. Otherwise we have that $i \geq 3$ and $k \geq j + 3$.

If i is odd, then we have that j is even and k is odd. Therefore, b_1, b_j, and b_k is a valid oriented triple, since $b_1 > b_i$ by our definition of listing black edges of a cycle, and $j - 1$, $k - j$, and $|C| + 1 - k$ are all odd.

If i is even, then we have that j is odd and k is even. If $b_2 > b_k$, then a transposition affecting b_2, b_j, and b_k is a 2-transposition, since these edges form a valid oriented triple, and $j - 2$, $k - j$, and $|C| + 2 - k$ are odd. If $b_2 < b_k$, we further divide the proof in the following cases.

- If $b_{k-1} > b_k > b_2$, then we have a 2-transposition acting on the valid oriented triple b_1, b_2, and b_{k-1}, since $b_1 > b_{k-1} > b_2$. Because k is even, we have that $(k - 1) - 2$ and $|C| + 1 - (k - 1)$ are odd, therefore, the three cycles created by this transposition are odd.
- If $b_{k-1} < b_k$, then we have a 2-transposition acting on the valid oriented triple b_i, b_k, and b_{k-1}, since $b_i > b_k > b_{k-1}$, and the created cycles have sizes $(k - 1) - i$, 1, and $|C| + i - k$, which are all odd.

Lemma 6. *If there is a valid oriented triple b_i, b_j, and b_k of an odd cycle $C = (b_1, b_2, \ldots, b_m)$ such that $i < j < k$ and $b_k > b_j > b_i$, then there is a valid oriented triple $b_{i'}$, $b_{j'}$, and $b_{k'}$ of C, with $i' < j' < k'$, such that $i' = 1$.*

Proof. Note that if $i = 1$ we can set $i' = i$, $j' = j$, and $k' = k$ as the valid oriented triple. Otherwise we have that $i \geq 2$.

If j is even, then we have a 2-transposition acting on the valid oriented triple b_1, b_j, and b_k, since in this case k is odd and $b_1 > b_k > b_j$. Otherwise, j is odd, and we have a 2-transposition acting on the valid oriented triple b_1, b_i, b_j, since i is even and $b_1 > b_j > b_i$.

Lemma 7. *If there is a valid oriented triple b_i, b_j, and b_k of an odd cycle $C = (b_1, b_2, \ldots, b_m)$, with $i < j < k$ and $b_j > b_i > b_k$, then there is a valid oriented triple $b_{i'}$, $b_{j'}$, and $b_{k'}$ of C, with $i' < j' < k'$, such that at least one of these conditions hold: (i) $i \in \{1, 2\}$; (ii) $j = i + 1$; or (iii) $k = j + 1$.*

Proof. Note that if $i \in \{1, 2\}$, or $j = i + 1$, or $k = j + 1$ we can set $i' = i$, $j' = j$, and $k' = k$ as the valid oriented triple. Otherwise we have that $i \geq 3$, $j \geq i + 3$ and $k \geq j + 3$.

If j is odd, then i and k are even. Consequently, we have a 2-transposition acting on the valid oriented triple b_1, b_i, and b_j (note that $b_1 > b_j > b_i$).

Since we have many cases to deal when j is even, we are just going to list the conditions and the corresponding valid oriented triple. From now on, we assume that j is even and both i and k are odd.

First we consider the black edge b_2:

- If $b_2 > b_j$, then b_2, b_i, and b_j is a valid oriented triple.
- If $b_2 < b_i$, then b_1, b_2, and b_i is a valid oriented triple.

Otherwise we have that $b_j > b_2 > b_i$. Now consider b_{i+1}, the black edge after b_i. Recall that $i + 1 < j$ and both $i + 1$ and j are even.

- If $b_{i+1} < b_k$, then b_1, b_{i+1}, and b_k is a valid oriented triple.
- If $b_{i+1} > b_2$, then b_i, b_{i+1}, and b_k is a valid oriented triple.
- If $b_2 > b_{i+1} > b_i$, then b_2, b_i, and b_{i+1} is a valid oriented triple.

Otherwise we have that $b_1 > b_j > b_2 > b_i > b_{i+1} > b_k$. Consider now b_{j+1}, the black edge right after b_j. Recall that $j + 1 < k$ and $j + 1$ is odd.

- If $b_{j+1} > b_i$, then b_1, b_{i+1}, and b_{j+1} is a valid oriented triple.
- Otherwise $b_{j+1} < b_i$, so b_i, b_j, and b_{j+1} is a valid oriented triple.

Lemmas 4–7 imply the following.

Corollary 1. *If there is a 2-transposition affecting an oriented cycle of $G(\pi)$, then there is a 2-transposition applied to black edges b_i, b_k, and b_k of a cycle $C \in G(\pi)$, with $i < j < k$, such that at least one of these conditions hold: (i) $i \in \{1, 2\}$; (ii) $j = i + 1$; or (iii) $k = j + 1$.*

Consider now Algorithm 1.

Lemma 8. *Given a permutation π, if there is a 2-transposition applied to $G(\pi)$, then Algorithm 1 finds a 2-transposition and returns it. Otherwise, the algorithm returns that there are no 2-transpositions for this permutation.*

Algorithm 1: Search for a 2-transposition

Input: A permutation π
Output: A 2-transposition τ, if it exists, or $\emptyset$

1 Construct $G(\pi)$
2 **if** *there is two even cycles in* $G(\pi)$ **then**
3 | **return** the 2-transposition from Lemma 3.2.5 of Christie [3]
4 **else**
5 | **for** *every oriented cycle* $C = (b_1, b_2, \ldots, b_m)$ *in* $G(\pi)$ **do**
6 | | **for** *every* $j \in \{2, \ldots, m-1\}$ **do**
7 | | | **for** *every* $k \in \{j+1, \ldots, m\}$ **do**
8 | | | | **if** $b_1 > b_k > b_j$ *and* $j-1$, $k-j$, *and* $m+1-k$ *are odd* **then**
9 | | | | | **return** the 2-transposition $\tau(b_j, b_k, b_1)$
10 | | | | **else if** $b_2 > b_k > b_j$ *and* $j-2$, $k-j$, *and* $m+2-k$ *are odd* **then**
11 | | | | | **return** the 2-transposition $\tau(b_j, b_k, b_2)$
12 | **for** *every oriented cycle* $C = (b_1, b_2, \ldots, b_m)$ *in* $G(\pi)$ **do**
13 | | **for** *every* $i \in \{3, \ldots, m-2\}$ **do**
14 | | | **for** *every* $j \in \{i+1, \ldots, m-1\}$ **do**
15 | | | | **if** $b_j > b_i > b_{j+1}$ *and both* $j-i$ *and* $m+i-j-1$ *are odd* **then**
16 | | | | | **return** the 2-transposition $\tau(b_{j+1}, b_i, b_j)$
17 | | | | **else if** $b_i > b_{j+1} > b_j$ *and both* $j-i$ *and* $m+i-j-1$ *are odd* **then**
18 | | | | | **return** the 2-transposition $\tau(b_j, b_{j+1}, b_i)$
19 | | | **for** *every* $k \in \{i+2, \ldots, m\}$ **do**
20 | | | | **if** $b_{i+1} > b_i > b_k$ *and both* $k-j-i$ *and* $m+i-k$ *are odd* **then**
21 | | | | | **return** the 2-transposition $\tau(b_k, b_i, b_{i+1})$
22 | **return** $\emptyset$ ▷ there are no 2-transpositions in $G(\pi)$

Proof. According to Lemma 3, the 2-transposition τ is applied to two even cycles or to an oriented cycle.

If there is a pair of even cycles in $G(\pi)$ the algorithm will always find a 2-transposition according to Lemma 3.2.5 of Christie [3].

If τ is applied to an oriented cycle $C = (b_1, b_2, \ldots, b_m)$, then there is a 2-transposition applied on a valid oriented triple $b_{i'}$, $b_{j'}$, and $b_{k'}$, with $i' < j' < k'$, such that $i' \in \{1, 2\}$ or $k' = j' + 1$ (Corollary 1). Since the algorithm exhaustive searches for all transpositions applied on edges b_i, b_j, and b_k, with $i < j < k$, such that $i \in \{1, 2\}$ or $k = j + 1$, we have that the algorithm will find a 2-transposition.

If there is no 2-transposition, then Algorithm 1 return no operation at its last line.

It takes linear time to construct $G(\pi)$ at line 1, and the complexity of lines 2 and 3 is also linear, as shown by Christie [3] in the proof of Lemma 3.2.5. The complexity of the search in lines 5–11 is the following, where c is a constant related to the operations of line 8 to 11:

$$\sum_{C \in G(\pi)} \sum_{j=2}^{|C|-1} \sum_{k=j+1}^{|C|} c = \sum_{C \in G(\pi)} \sum_{j=2}^{|C|-1} (|C|-j)c < c \sum_{C \in G(\pi)} |C|^2 = O(n^2),$$

since $\sum_{C \in G(\pi)} |C| \leq n+1$.

The complexity of lines 12–21 can be shown to be $O(n^2)$ in a similar way, so it follows that Algorithm 1 has a time complexity of $O(n^2)$.

4 $O(n^5)$ Time 1.375-Approximation Algorithm

In this section we show how to achieve the $O(n^5)$ time complexity while guaranteeing the 1.375 approximation factor. We first recall that Algorithm 1 cannot be used to list all possible 2-transpositions of a permutation, although it always returns a 2-transposition if $G(\pi)$ admits a 2-transposition.

Consider Algorithm 3, which uses Algorithm 2 to find a sequence of two 2-transpositions, if it exists, and EH_Algorithm, the original 1.375-approximation algorithm from Elias and Hartman [7].

Algorithm 2: Find a sequence of two 2-transpositions, if it exists

Input: A permutation π
Output: A sequence of two 2-transposition, if it exists
1 Construct $G(\pi)$
2 Let z be the number of black edges in $G(\pi)$
3 **for** *every* $i \in \{1, \ldots, z-2\}$ **do**
4   **for** *every* $j \in \{i+1, \ldots, z-1\}$ **do**
5   **for** *every* $k \in \{j+1, \ldots, z\}$ **do**
6   **if** $\tau(i,j,k)$ *is a 2-transposition* **then**
7          $\pi' \leftarrow \pi \cdot \tau$
8          $\tau' \leftarrow$ Algorithm_1(π')
9   **if** $\tau' \neq \emptyset$ **then**
10   **return** (τ, τ')
11 **return** $(\emptyset, \emptyset)$ ▷ there is no sequence of two 2-transpositions

---

Algorithm 3: An improved 1.375-approximation algorithm for **SbT**

Input: A permutation π
Output: A sequence of transpositions $\tau_1, \ldots, \tau_r$ that sorts π
1 $(\tau_1, \tau_2) \leftarrow$ Algorithm_2(π)
2 **if** $\tau_1 \neq \emptyset$ **then**
3    $\pi' \leftarrow \pi \cdot \tau_1 \cdot \tau_2$ ▷ a sequence of two 2-transpositions exists
4   **return** (τ_1, τ_2) + EH_Algorithm(π')
5 **else**
6   **return** EH_Algorithm(π)

Recall that the main problem of the EH_Algorithm (and other proposed algorithms based on it) to guarantee the 1.375 approximation factor is the possible overlook of a sequence of two 2-transpositions after transforming π into a simple permutation π'.

We will first use the Algorithm 2 that searches for a sequence of two 2-transpositions for π. For each τ generated using all possible values for i, j, and k, if τ is a 2-transposition the algorithm will apply this transposition and call Algorithm 1 with the resulting permutation; if Algorithm 1 returns a second transposition τ', then the algorithm returns this pair of 2-transpositions at line 10. If no pair of 2-transpositions exists, the algorithm returns an empty sequence.

Note that line 6 takes $O(1)$ time, line 7 takes $O(n)$ time, and line 8 uses Algorithm 1 that takes $O(n^2)$ time. This algorithm also uses nested loops at lines 3–5 to search for the first 2-transposition $\tau(i, j, k)$ using all possible indices for i, j, and k. Since lines 6–10 are inside three nested loops, it follows that Algorithm 2 executes in $O(n^5)$ time.

After using Algorithm 2 once at line 1, the Algorithm 3 uses the EH_Algorithm either at line 4 or at line 6 on the resulting permutation. Since the EH_Algorithm takes $O(n^2)$ time, it follows that Algorithm 3 has a time complexity of $O(n^5)$.

Since Algorithm 3 guarantees to first apply a sequence and 2-transpositions for π if it exists, and only after that uses EH_Algorithm that transforms the input permutation into a simple permutation, it is guaranteed that the approximation factor of Algorithm 3 is also 1.375.

5 Experimental Analysis

In this section, we present the practical results of Algorithm 3. In addition, we compare it to the results provided by the approximation algorithms presented by Elias and Hartman [7] and Silva *et al.* [9] which we will refer to as **EH** and **SKRW**, respectively.

We tested Algorithm 3 in all permutations π of size $m \leq 12$ and compared the results with their exact distances using the GRAAu tool [8]. We will refer to Algorithm 3 simply as **ALG3**.

Table 1 summarizes the results provided by the algorithms **ALG3**, **EH**, and **SKRW**. Columns **MAX APPROX** and **AVG APPROX** represent, for each permutation size group, the maximum and average approximation ratio observed, respectively. Columns **AVG DIST** and **% OPT SOLUTIONS** show, for each permutation group, the information regarding the average distance and the percentage of instances such that an optimal solution was reached.

The **EH** algorithm returned a sequence with an approximation above 1.375 (compared to the exact distance) in 2 instances of size 8, 20 instances of size 9, 110 instances of size 10, 440 instances of size 11, and 1448 instances of size 12. On the other hand, algorithms **SKRW** and **ALG3** did not return any sequence with an approximation above 1.333.

Comparing the maximum approximation ratio observed by algorithms **ALG3** and **SKRW** in Table 1, it is possible to note that, for all the permutation size groups, the value was the same and not exceeded the ratio of 1.333. Looking

Table 1. Practical results of algorithm **ALG3** compared to algorithms **EH** and **SKRW** in all permutations of size up to 12, excluding the identity permutations ι^n.

	MAX APPROX			AVG APPROX			AVG DIST			% OPT SOLUTIONS		
n	EH	SKRW	ALG3	EH	SKRW	ALG3	EH	SKRW	ALG3	EH	SKRW	ALG3
2	1.00	1.00	1.00	1.00	1.00	1.00	1.00	1.00	1.00	100.00	100.00	100.00
3	1.00	1.00	1.00	1.00	1.00	1.00	1.20	1.20	1.20	100.00	100.00	100.00
4	1.00	1.00	1.00	1.00	1.00	1.00	1.6086	1.6086	1.6086	100.00	100.00	100.00
5	1.00	1.00	1.00	1.00	1.00	1.00	2.0924	2.0924	2.0924	100.00	100.00	100.00
6	1.33	1.00	1.00	1.0004	1.00	1.00	2.6063	2.6050	2.6050	99.86	100.00	100.00
7	1.33	1.25	1.25	1.0129	1.0113	1.0014	3.1762	3.1704	3.1311	94.90	95.47	99.40
8	1.50	1.25	1.25	1.0210	1.0183	1.0042	3.7178	3.7076	3.6512	91.64	92.65	98.29
9	1.50	1.25	1.25	1.0301	1.0256	1.0085	4.2796	4.2603	4.1846	86.62	88.54	96.10
10	1.50	1.25	1.25	1.0341	1.0282	1.0125	4.8051	4.7772	4.7032	83.80	86.53	93.94
11	1.50	1.33	1.33	1.0392	1.0321	1.0170	5.3526	5.3157	5.2367	79.40	82.98	90.88
12	1.50	1.33	1.33	1.0415	1.0336	1.0206	5.8694	5.8248	5.7514	76.67	80.91	88.27

at the average approximation ratio values provided by algorithms **ALG3** and **SKRW**, we can note that in the groups with sizes from 2 up to 6, the record 1.0 was maintained, which means that, for all the instances of these groups, both algorithms provided an optimal solution (**% OPT SOLUTIONS** column also shows this information). Besides, the **ALG3** algorithm provided better results considering the metrics of average approximation, average distance, and percentage of optimal solutions reached in the permutation size groups greater than 6.

It is important to note that the percentage of instances in which the **ALG3** algorithm provides an optimal solution was greater than 88% for all permutation size groups. Algorithm **SKRW** maintains this behavior only for groups of permutations with a size less or equal to 9. This behavior indicates that **ALG3** not only brings an improvement considering the complexity but also from the practical perspective.

6   Conclusion

In this work, we show a new algorithm to find a sequence of two 2-transpositions in $O(n^5)$ time, decreasing the time complexity of $O(n^6)$ of a recently published algorithm that corrected an issue in the 1.375 approximation algorithm from Elias and Hartman.

We tested using this new algorithm before using the algorithm from Elias and Hartman and showed that in the set of permutations of sizes up to 12, this combination returned the exact distance in more than 90% of the cases, while the other two algorithms alone returned the exact distance in less than 85% of all instances each.

Acknowledgment. This work was supported by the National Council of Technological and Scientific Development, CNPq (grants 140272/2020-8, 202292/2020-7, and 425340/2016-3), the Coordenação de Aperfeiçoamento de Pessoal de Nível Superior -

Brasil (CAPES) - Finance Code 001, and the São Paulo Research Foundation, FAPESP (grants 2013/08293-7, 2015/11937-9, and 2019/27331-3).

References

1. Bafna, V., Pevzner, P.A.: Sorting by transpositions. SIAM J. Discret. Math. **11**(2), 224–240 (1998)
2. Bulteau, L., Fertin, G., Rusu, I.: Sorting by transpositions is difficult. SIAM J. Discret. Math. **26**(3), 1148–1180 (2012)
3. Christie, D.A.: Genome Rearrangement Problems. Ph.D. thesis, Department of Computing Science, University of Glasgow (1998)
4. Dias, U., Dias, Z.: An improved 1.375-approximation algorithm for the transposition distance problem. In: Proceeding of the 1st ACM International Conference on Bioinformatics and Computational Biology (BCB'2010), pp. 334–337. ACM, New York (2010)
5. Dias, U., Dias, Z.: Extending bafna-pevzner algorithm. In: Proceedings of the 1st International Symposium on Biocomputing (ISB'2010), pp. 1–8. ACM, New York (2010)
6. Dias, U.M.: Problemas de Comparação de Genomas. Ph.D. thesis, Institute of Computing, University of Campinas (2012). In Portuguese
7. Elias, I., Hartman, T.: A 1.375-approximation algorithm for sorting by transpositions. IEEE ACM Trans. Comput. Biol. Bioinf. **3**(4), 369–379 (2006)
8. Galvão, G.R., Dias, Z.: An audit tool for genome rearrangement algorithms. J. Exp. Algorithmics **19**, 1–34 (2014)
9. Silva, L.A.G., Kowada, L.A.B., Rocco, N.R., Walter, M.E.M.T.: A new 1.375-approximation algorithm for sorting by transpositions. Algorithms Mol. Biol. **17**(1), 1–17 (2022). https://doi.org/10.1186/s13015-022-00205-z
10. Walter, M.E.M.T., Dias, Z., Meidanis, J.: A new approach for approximating the transposition distance. In: Titsworth, F.M. (ed.) Proceedings of the 7th String Processing and Information Retrieval (SPIRE'2000), pp. 199–208. IEEE Computer Society, Los Alamitos (2000)

In Silico Analysis of the Genomic Potential for the Production of Specialized Metabolites of Ten Strains of the Bacillales Order Isolated from the Soil of the Federal District, Brazil

Felipe de Araújo Mesquita[1], Waldeyr Mendes Cordeiro da Silva[1,2], and Marlene Teixeira De-Souza[1(✉)]

[1] Depto. Biologia Celular, Instituto de Ciências Biológicas, Universidade de Brasília (UnB), Federal, Brazil
felipedearaujomesquita@gmail.com, waldeyr.mendes@ifg.edu.br, marlts@unb.br
[2] Instituto Federal de Goiás (IFG), Formosa, Goiás, Brazil

Abstract. Secondary or specialized metabolites play an important ecological role for the producing organisms. Bacteria isolated from soils are a major source of specialized metabolites. Species of Bacillus and related genera, collectively referred to as aerobic endospore-forming bacteria (AEFB), produce specialized metabolites with high structural and functional diversity. In this study, ten genomes of AEFB strains isolated from the soil of Federal District, Brazil, were scanned for specialized metabolism genes. Using the antiSMASH 6.0 bacterial standalone version, we identified 153 putative gene clusters codifying for specialized metabolite synthesis in these ten strains. Such clusters encode, for example, enzymes for bacillibactin, bacillisin, macrolactin H, bacillaene, paenibacterin, nostamid A and macrobervin, revealing pathways 100% similar to the genomic information available in the antiSMASH database. The results suggest that AEFB are promising for exploring known and unknown specialized metabolites, notably antimicrobial agents.

Keywords: AEFB · Bacillus · Secondary metabolites · Bacteria

1 Introduction

Secondary or specialized metabolites play an essential ecological role in producing organisms such as plants, fungi, and bacteria and are often explored for commercial and technological applications [7]. Although not essential for the primary metabolism, these molecules can help to guarantee the availability of nutrients in a competitive environment [16] as the soil provides a rich matrix with a great diversity of functional bioproducts, despite being a nutrient-limited environment [14,19]. Several classes of specialized metabolites, such as

N. M. Scherer and R. C. de Melo-Minardi (Eds.): BSB 2022, LNBI 13523, pp. 158–163, 2022.
https://doi.org/10.1007/978-3-031-21175-1_17

polyketides (PK), non-ribosomal peptides (NRP), alkaloids, and terpenes, are synthesized by three biosynthetic pathways. The species of Bacillus and other related genera - collectively called aerobic endospore-forming bacteria (AEFB) - produce high structural and functional variability of specialized metabolites and have the soil as the main reservoir [8]. Although many types of specialized metabolites have already been isolated and identified, there is a consensus that many of these molecules have not yet been described in microorganisms [17]. We could emphasize NRP with antimicrobial activity, including lipopeptides, cyclic peptides, and in some cases, bacteriocins [6,17], which are currently promising alternatives for a new generation of antimicrobials. AEFB have a high potential for exploiting these peptides [17]. Thus, studies to prospect the existence of ways to produce specialized metabolites in these bacteria are essential to aid in combating the historic and growing wave of resistance. The description of specialized metabolite classes such as terpenes and alkaloids is still scarce for AEFB. Analysis for identifying possible clusters involved in synthesizing these types of metabolites specialized in AEFB strains is also relevant because the production of these compounds has already been identified in *B. subtilis* and *B. megaterium* strains [13]. Based on the elements presented, the following questions arise: what is the potential genome for producing specialized metabolites in AEFB strains? What are the main classes of these metabolites in this group of microorganisms? Recent *in silico* tools, like antibiotics and secondary metabolite analysis shell - antiSMASH [3], have allowed us to search for clusters of genes encoding metabolic pathways devoted to synthesizing specialized metabolites. antiSMASH can detect gene clusters known as biosynthetic gene clusters (BGC), involved in the synthesis strategies of these products in plants, fungi, and bacteria. Based on this scenario and the genome's availability of ten AEFB species isolated, sequenced, assembled, and annotated by our laboratory, we aimed to verify them in silico, seeking to identify BGC involved in specialized metabolism.

2 Material and Methods

2.1 Genomic Sequences

The ten strains used in this work were isolated from the Federal District, Brazil soil, and stored as described in [1]. Total DNA from the ten strains was obtained using the Wizard Genomic DNA Purification Kit (Promega), according to the manufacturer's instructions [5]. The purified total DNA was sequenced using the Illumina MiSeq platform, and the genomes were deposited on the NCBI (Table 1).

2.2 Analysis of the Genomic Potential for the Production of Specialized Metabolites

The genomic sequences were downloaded directly from the NCBI platform to analyze the genomic potential of the ten strains used in this work. We ran anti-SMASH [3] over these genomes to identify biosynthetic gene clusters (BGC)

Table 1. Genomes of AEFB strains used in this work.

AEFB strain	Strain ID	NCBI ID	Access level
Lysinibacillus fusiformis	SDF0005	SAMN12262387	Restrict
Bacillus pumilus	SDF0011	SAMN12262389	Restrict
Bacillus oleronius	SDF0015	SAMN12262388	Restrict
Bacillus simplex	SDF0024	SAMN12262386	Restrict
Bacillus velezensis	SDF0141	SAMN12262409	Restrict
Bacillus velezensis	SDF0150	SAMN12262428	Restrict
Paenibacillus sp.	SDF0016	SAMN06921004	Public
Paenibacillus sp.	SDF0028	SAMN06917517	Public
Lysinibacillus sp.	SDF0037	SAMN06921005	Public
Lysinibacillus sp.	SDF0063	SAMN06917521	Public

related to specialized metabolism. The parameter used for the accuracy of detecting clusters was relaxed with algorithms provided by antiSMASH (KnownClusterBlast, ActiveSiteFinder, ClusterPfam, ClusterBlast and Pfam-based GO term annotation). The collected results were tabulated to identify the main classes of clusters related to specialized metabolism and identify potential metabolites produced by the strains. The most promising results were extracted using in-house Python scripts for future analysis.

3   Results

Results revealed 153 putative biosynthetic gene clusters (BGC) in the ten genomes analyzed using the antiSMASH. Among these, 20 different groups of specialized metabolites were detected (Fig. 1), the NRPS (non-ribosomal peptide synthase) type being the most numerous with 47 (30.7%) BGC recognized; followed by trans AT-PKS (polyketide synthase acyltransferase) with 17 (11.1%); terpenes with 16 (10.5%); T3PKS (type III polyketide synthase) with 12 (7.8%); betalactone with 10 (6.5%); RiPP-like (ribosomal peptides modified after translation) with 7 (4.5%). PKS I (polyketide synthase type I) and siderophores comprised 6 (4%) of the BGC detected.

Among the ten SDF strains analyzed, the genome of *Paenibacillus sp.* SDF0028 showed the highest number of recognized BGC with 38 (24.8%) clusters identified, followed by *B. velezensis* SDF0150 with 22 (14.3%) and *B. velezensis* SDF0141 with 21 (13.7%); Still, *Lysinibacillus sp.* with SDF0063, 19 (12.4%); *B. pumilus* SDF0011 with 16 (10.4%); *B. simplex* SDF0024 and *Paenibacillus sp.* SDF0016 with 10 (6.5%), *Lysinibacillus sp.* SDF0037 with 9 (5.8%), *L. fusiformis* SDF0005, 5 (3.2%) and *B. oleronius* SDF0015 with 3 (1.9%).

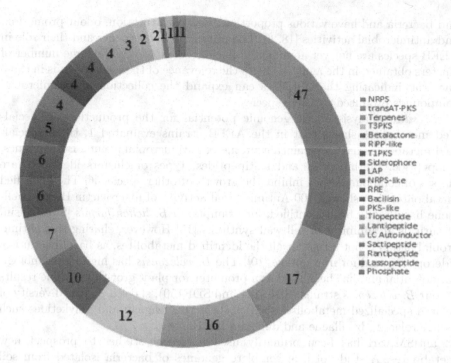

Fig. 1. Groups and number of specialized metabolites recognized in the ten genomes of SDF strains of interest using the antiSMASH tool (clockwise list).

Among the 153 recognized BGC, 63 (41.1%) clusters related to producing potentially specialized metabolites were identified. Eleven BGC (7.1%) in the strains *B. velezensis* SDF0141 and *B. velezensis* SDF0150, showed 100% similarity with the antiSMASH data for BGC involved in the production of bacillibactin, bacilisin, macrolactin H and bacillaene; The same occurred for nostamide A and paenibacterin in the *Paenibacillus sp.* SDF0028 and macrobervin in the *Lysinibacillus sp.* SDF0063.

4 Discussion

The biosynthetic gene clusters (BGC) identification on these SDF strains represents an in silico finding of their genomic potential for specialized metabolism. NRPS, the most numerous clusters identified in this study, synthesizes non-ribosomal peptides whose activities encompass antifungal, antibacterial, antiviral, and immunosuppressive properties [5]. NRPS use many substrates from amino acids, fatty acids and carboxylic acids due to the structural and topological variety identified in these catalysts [5]. This diversity explains the number of NRPS gene clusters identified in this study.

Terpenes derive from the condensation of isoprene units and are fundamental constituents of essential oils. These compounds are produced by plants, fungi, and bacteria and have various properties, from pigmentation, odour promotion, and antimicrobial activities [18, 20] The production of terpenes and their role in AEFB species are not yet abundantly elucidated. However, the large number of clusters obtained in the analysis shows the relevance of these compounds in these bacteria, indicating that terpenes can expand the collection of antimicrobial compounds produced by these species.

In silico analysis of the genomic potential for the production of specialized metabolites shows that in the AEFB strains evaluated 1, these specialized metabolites have a significant profile of antimicrobial agents. Betalactones, thiopeptide, ranthipeptide and lantipeptides, types of clusters identified, are classes of compounds that inhibit the growth of other species [6] The identified metabolites that showed 100 Antimicrobial activity of macrolactin H and bacillaene has already been identified, for example, in *B. licheniformis* [2] Bacilisin, in turn, is an inhibitor of cell wall synthesis [15] However, sharing this antimicrobial action is not a rule for all the identified metabolites, as bacillibactin is a siderophore used for iron uptake [10]. The *B. velezensis* has high biotechnological potential and has been used as a promoter for plant growth [11]. The results on our *B. velezensis* strains (SDF0141 and SDF150), showed a great diversity of related specialized metabolites such as surfactin, fengicin and polyketides such as macrolactin, bacillaene and difficidin [11, 15].

antiSMASH has been primarily used in recent studies to prospect new metabolites. A study of near-complete genomes of bacteria isolated from soil and analyzed them using antiSMASH showed that members from *Acidobacteria*, *Verrucomicobia* and *Gemmatimonadetes* were found to encode diverse PK and NRP BGC that were thought to have diverged from well-studied gene clusters [4]. Besides soil, marine microorganisms such as *Roseobacter* and *Pseudovibrio* genera were related as a potential and largely untapped source of specialized metabolites [9, 12].

In summary, although some gene clusters have high similarity rates, it does not rigorously indicate that the identified clusters effectively synthesize the putative metabolites. However, the high similarity index is a clue for their genomic potential. Thus, the clusters identified with a low similarity index and those that did not obtain a correspondence possibly indicate a great diversity of not yet described metabolites, reinforcing the need for future studies.

References

1. de Andrade Cavalcante, D., et al.: Ultrastructural analysis of spores from diverse Bacillales species isolated from Brazilian soil. Environ. Microbiol. Rep. **11**(2), 155–164 (2019)
2. Arbsuwan, N., et al.: Purification and characterization of macrolactins and amicoumacins from bacillus licheniformis bfp011: a new source of food antimicrobial substances. CyTA-J. Food **16**(1), 50–60 (2018)
3. Blin, K., et al.: antiSMASH 5.0: updates to the secondary metabolite genome mining pipeline. Nucleic Acids Res. **47**(W1), W81–W87 (2019)

4. Crits-Christoph, A., Diamond, S., Butterfield, C.N., Thomas, B.C., Banfield, J.F.: Novel soil bacteria possess diverse genes for secondary metabolite biosynthesis. Nature **558**(7710), 440–444 (2018)
5. Guzmán-Chávez, F., Zwahlen, R.D., Bovenberg, R.A., Driessen, A.J.: Engineering of the filamentous fungus penicillium chrysogenum as cell factory for natural products. Front. Microbiol. **9**, 2768 (2018)
6. Heilbronner, S., Krismer, B., Brötz-Oesterhelt, H., Peschel, A.: The microbiome-shaping roles of bacteriocins. Nat. Rev. Microbiol. **19**(11), 726–739 (2021)
7. Keller, N.P., Turner, G., Bennett, J.W.: Fungal secondary metabolism-from biochemistry to genomics. Nat. Rev. Microbiol. **3**(12), 937–947 (2005)
8. Mandic-Mulec, I., Prosser, J.I.: Diversity of endospore-forming bacteria in soil: characterization and driving mechanisms. In: Logan, N., Vos, P. (eds.) Endospore-Forming Soil Bacteria, pp. 31–59. Springer, Berlin (2011). https://doi.org/10.1007/978-3-642-19577-8_2
9. Martens, T., et al.: Bacteria of the roseobacter clade show potential for secondary metabolite production. Microb. Ecol. **54**(1), 31–42 (2007). https://doi.org/10.1007/s00248-006-9165-2
10. Miethke, M., Bisseret, P., Beckering, C.L., Vignard, D., Eustache, J., Marahiel, M.A.: Inhibition of aryl acid adenylation domains involved in bacterial siderophore synthesis. FEBS J. **273**(2), 409–419 (2006)
11. Rabbee, M.F., Ali, M.S., Choi, J., Hwang, B.S., Jeong, S.C., Baek, K.H.: Bacillus velezensis: a valuable member of bioactive molecules within plant microbiomes. Molecules **24**(6), 1046 (2019)
12. Romano, S.: Ecology and biotechnological potential of bacteria belonging to the genus pseudovibrio. Appl. Environ. Microbiol. **84**(8), e02516-17 (2018)
13. Sato, T., Yoshida, S., Hoshino, H., Tanno, M., Nakajima, M., Hoshino, T.: Sesquarterpenes (c35 terpenes) biosynthesized via the cyclization of a linear c35 isoprenoid by a tetraprenyl-β-curcumene synthase and a tetraprenyl-β-curcumene cyclase: identification of a new terpene cyclase. J. Am. Chem. Soc. **133**(25), 9734–9737 (2011)
14. Sharrar, A.M., Crits-Christoph, A., Méheust, R., Diamond, S., Starr, E.P., Banfield, J.F.: Bacterial secondary metabolite biosynthetic potential in soil varies with phylum, depth, and vegetation type. MBio **11**(3), e00416-20 (2020)
15. Shenderov, B.A., Sinitsa, A.V., Zakharchenko, M.M., Lang, C.: Cellular metabiotics and metabolite metabiotics. In: METABIOTICS, pp. 63–75. Springer, Cham (2020). https://doi.org/10.1007/978-3-030-34167-1_14
16. Singh, B.P., Rateb, M.E., Rodriguez-Couto, S., Polizeli, M.D.L.T.D.M., Li, W.J.: Microbial secondary metabolites: recent developments and technological challenges. Front. Microbiol. **10**, 914 (2019)
17. Sumi, C.D., Yang, B.W., Yeo, I.C., Hahm, Y.T.: Antimicrobial peptides of the genus bacillus: a new era for antibiotics. Can. J. Microbiol. **61**(2), 93–103 (2015)
18. Teramoto, J.R.S., Sachs, R.C.C., Garcia, V.L.: Atividade antimicrobiana das folhas de duas variedades de oliveira e a contextualização deste coproduto da produção paulista e mundial de azeite de oliva. Rev. Intellectus **37**, 63–83 (2017)
19. Tyc, O., Song, C., Dickschat, J.S., Vos, M., Garbeva, P.: The ecological role of volatile and soluble secondary metabolites produced by soil bacteria. Trends Microbiol. **25**(4), 280–292 (2017)
20. Valduga, E., Tatsch, P.O., Tiggemann, L., Treichel, H., Toniazzo, G., Zeni, J., Di Luccio, M., Furigo, A.: Produção de carotenoides: microrganismos como fonte de pigmentos naturais. Quim. Nova **32**(9), 2429–2436 (2009)

Author Index

Printed in the United States
by Baker & Taylor Publisher Services